Design of Jigs, Fixtures
and
Press Tools

Design of Jigs, Fixtures and Press Tools

K. Venkataraman

External Faculty
Mechanical Engineering Department
Anna University,
Chennai, India

WILEY

John Wiley & Sons Ltd.

Athena
ACADEMIC

Design of Jigs, Fixtures and Press Tools

K. Venkataraman

This Edition Published by

John Wiley & Sons Ltd
The Atrium, Southern Gate
Chichester, West Sussex
PO19 8SQ United Kingdom
Tel : +44 (0)1243 779777
Fax : +44 (0)1243 775878
e-mail : customer@wiley.com
Web : www.wiley.com

For distribution in rest of the world other than the Indian sub-continent and Africa.

Under licence from:

Athena Academic Ltd
Suite LP24700, Lower Ground Floor
145-157 St. John Street,
London
ECIV 4PW.
United Kingdom
e-mail: athenaacademic@gmail.com
Web: www.athenaacademic.com

ISBN : 978-11-1915-567-6

Library Congress Cataloging-in-Publication Data

A catalogue record for this book is available from the British Library.

To
My Dear Parents
Late V. Kalyanaraman
and
Mrs. Rajalakshmi Kalyanaraman

Contents

PART II: PRESS TOOLS

Foreword

The book, Design of Jigs, Fixtures and Press Tools by K. Venkataraman, is intended for undergraduate students in Mechanical Engineering and Production/Manufacturing Engineering students. It is a treatise on two major topics in 'Tooling', *viz.* (*a*) Jigs and Fixtures and (*b*) Press Tools and thus makes it comprehensive for undergraduate students of Mechanical Engineering and allied branches. The book covers all the major topics in the subject. Some of the salient features of the book are as follows:

- Exhaustive illustrations covering almost all variants in the subject of Jigs, Fixtures and Press Tools.
- An appendix at the end of Part I of the book dealing with the mechanics of cutting tool operation and the forces involved in various tools such as turning, milling, drilling and broaching.
- An appendix on worked examples for the first part showing the 2-D drawings of the typical jigs/fixtures as well as a 3-D model of a jig will be very useful for the beginner.
- 3-D models of fixtures such as (*a*) a common vise used in milling operations, (*b*) Three-jaw chuck and in the field of press tools a model of a progressive die with associated components for making the students understand the concepts better.
- The final chapter in Part II showing typical worked examples of drawing dies.
- Separate appendices giving suggested questions and answers in both the parts to facilitate review of the subject by the students.

I am sure that this book will go a long way in filling the long-felt gap by covering both the topics of tooling under one cover. I congratulate the author for this effort and hope the students make full use of it.

Dr. K. Srinivasan
Director
AU-FRG Institute for CAD/CAM
Anna University

Preface

The subject 'Tooling' encompasses areas such as (*i*) Cutting Tools (*ii*) Gauges, (*iii*) Jigs and Fixtures, and (*iv*) Press Tools. Each of these fields is very vast. To become a successful professional, be it a designer or a production or manufacturing engineer, the student needs to have an in-depth knowledge of all the above topics. In addition to the expertise needed in such specific areas, knowledge of Materials Science, Costing and Economics and Computer Modeling of components and sub-systems is also essential.

The present book, Design of Jigs, Fixtures and Press Tools, is aimed at providing the introductory knowledge on the subject to the undergraduate students of mechanical and manufacturing engineering of Anna University. Many of the universities in India prescribe a syllabus that contains both Design of Jigs and Fixtures, and Design of Press Tools in a single semester course. Keeping the above in mind, this book is designed in two parts. Part I deals with Jigs and Fixtures and Part II is earmarked exclusively for the study of Press Tools. Both these subjects are built progressively in successive chapters. A separate appendix in each part, provides short answer questions with answers, which will help the students in clarifying doubts and strengthen their knowledge base. The explanatory notes and illustrations provided in the book will serve the purpose of awakening the interest of the students and invoking in them the passion for tooling in their study of mechanical, manufacturing, or production engineering.

Finally, I wish to express my gratitude to the Anna University for providing support in my endeavour to write a book on the subject.

K. Venkataraman

PART-I

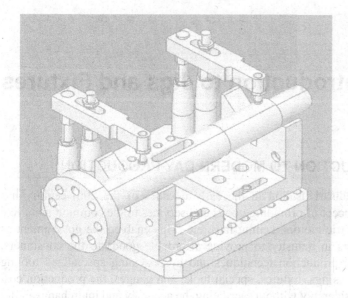

JIGS AND FIXTURES

Introduction to Jigs and Fixtures

1.1 INTRODUCTION TO MODERN DAY PRODUCTION

The advent of industrialisation in the early decades of the 20th century has ushered in the concept of providing goods and services to the common man, like the motorcar, electric motors, ceiling fans, etc. This enabled the government as well as the leaders in industry to provide affordable goods to the consumers. By improving the production techniques and by providing specialized tooling equipments, such as jigs, fixtures, special tools, and gauges, the production cost have reduced considerably without sacrificing the accuracy and interchangeability of parts and components. To achieve the desired quality and quantity of production, the concept of accuracy and interchangeability go hand in hand. They play a major role in meeting the present day classes of engineering production, namely, "flow production" and "batch production".

To necessitate the need of jigs, fixtures and special tools, the four main engineering classes of production are as follows:

1. Job Production: This involves the manufacture of specialized components or systems to meet the specific needs of the customers. Examples of job production are the manufacture of jigs, fixtures and press tools.

2. Batch Production: Some of the examples of batch production are the manufacture of aeroplanes, aero-engines, battle tanks, etc., that use the concept of intermittent manufacture of large range of products, produced in batches. Some brands of motorcars like "Benz" and "BMW" may be classified under "batch production" as they are required to meet specific requirements.

3. Flow Production: In flow production, the standardised finished products are produced in plants, specifically laid out for this purpose. Examples of flow production are the modern motorcar plants.

4. Mass Production: In this type of plants, the products are produced in mass quantities by specialised and repetitive methods, without requiring specialised layouts as in the case of flow production. Examples are mass production of screws, pins, hand tools, like chisels, spanners, hammers, etc.

Design of Jigs, Fixtures and Press Tools, First Edition. K. Venkataraman.
© K. Venkataraman 2015. Published by Athena Academic Ltd and John Wiley & Sons Ltd.

1.2 DEFINITION OF JIGS, FIXTURES AND TOOLING

As explained earlier the present day trend is to produce components and systems to meet the basic specifications of:

    (*i*) Accuracy

    (*ii*) Interchangeability

    (*iii*) Economic production rate.

    In order to achieve the above objectives the following tooling equipments are deployed:

    (*i*) Jigs

    (*ii*) Fixtures

    (*iii*) Special tools like broaching tool, gear shaping tools, special class of taps and reamers

    (*iv*) Gauges to verify if dimensions are within the limits.

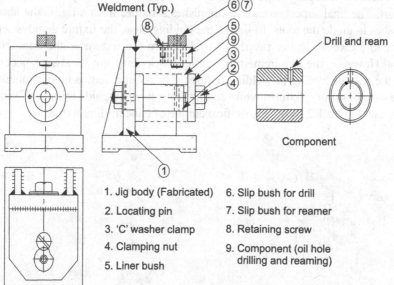

1. Jig body (Fabricated)
2. Locating pin
3. 'C' washer clamp
4. Clamping nut
5. Liner bush
6. Slip bush for drill
7. Slip bush for reamer
8. Retaining screw
9. Component (oil hole drilling and reaming)

Fig. 1.1 Example of a Drilling Jig

    A jig is a device, in which the component is clamped in a specific location so that cutting tools are guided to perform one or more operations. Jigs, which are independent devices, are fastened to the table of a machine tool. They are so designed to facilitate loading and unloading of components with ease. The third feature of a jig is that it has locating devices to position a component in a unique way. The fourth aspect of a jig specification is the gripping of the workpiece through a clamping device. There are elaborate methods to clamp, namely, (*a*) threaded fasteners, (*b*) cam clamps, (*c*) 'V' type sliding clamps, (*d*) pneumatic clamps, (*e*) hydraulic clamps, etc. An exclusive chapter is earmarked in later part of this book, which deals with various clamping techniques. The fifth aspect of the specification of a jig, which distinguishes the same from a fixture, is the guiding bushes which are fixed/fastened to the jig body or frame and act as guides to the tools, especially drilling, reaming or face milling cutters. They enable the tool to be positioned exactly with respect to the component and more precisely in relation to the location of the hole to be drilled, or the hole to be reamed. A later chapter deals with various types of bushes employed, such as plain bush, collared bush and renewable bush. Figure 1.1 gives a specific example of a drilling jig.

A fixture is also a device, which is fastened to the table of a machine tool, such as milling machine, and in the case of turning operation in a lathe, the fixture is fastened to the chuck or a faceplate. The device also enables loading and unloading of components with ease. The third aspect of unique location of workpiece in relation to the fixture also holds good as in the case of a jig. The fourth aspect of clamping is given more emphasis in fixture design as the clamping force should be able to withstand the cutting force which may not be along with the direction of gravity, and hence needs to be analysed more closely. At the same time, the clamping force should not be excessive, as it may cause damage to the part. The final aspect, which distinguishes a fixture from a jig, is the absence of bushes to guide the tools. In lieu of the guiding bush, the fixture deploys setting blocks to locate the cutter properly in relation to the fixture or the components *per se*. However, the requirement of setting blocks may not be always necessary as in the case of turning or welding fixtures. The requirement is more pronounced in the case of milling fixtures while cutting slots, keyways, side milling of fastener heads, etc. Figure 1.2 shows a specific example of gang milling fixture.

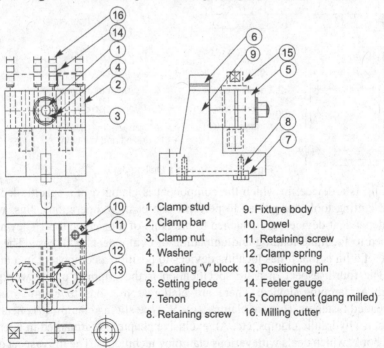

1. Clamp stud	9. Fixture body
2. Clamp bar	10. Dowel
3. Clamp nut	11. Retaining screw
4. Washer	12. Clamp spring
5. Locating 'V' block	13. Positioning pin
6. Setting piece	14. Feeler gauge
7. Tenon	15. Component (gang milled)
8. Retaining screw	16. Milling cutter

Fig. 1.2 Example of Gang Milling Fixture

1.3 FUNDAMENTAL CONCEPTS IN THE DESIGN OF JIGS AND FIXTURES

The basic difference between the design of "Jigs and Fixtures" and that of machine tool components is that the designing of jigs and fixtures calls for extreme accuracy followed by rigidity, whereas in the case of various machine elements, the concept of stress analysis plays a vital role. Therefore, the design of such special devices calls for in-depth knowledge on material specifications, mechanics of metal cutting, concepts of accuracy, simplicity, strength, safety and economy.

Hence, the designer of these special tools should be capable of preparing manufacturing drawings to meet the specific requirements of each job, its production scheme, rate of production, and the level of dynamic forces involved. As indicated earlier, the design of jig has the following aspects or elements:

(*i*) Unique location of components with respect to the jig

(*ii*) Ease of loading and unloading the components

(*iii*) Clamping of the components so as to impart adequate clamping force and also to have ease in operation

(*iv*) Guiding the cutting tools

(*v*) Provision for swarf removal

(*vi*) Proper fastening methods to hold the jig to the table (in the case of radial drilling machines)

(*vii*) Holding the assembly together so as to withstand the cutting forces which occur at frequent intervals causing static and dynamic forces

(*viii*) Provision for replacement of bushes, in case different tools like reaming subsequent to drilling are used for drilling different diameter holes in the same location.

Following are the major elements in the design of fixtures:

(*i*) Unique location of components with respect to the fixture

(*ii*) Clamping techniques to be adopted to deploy adequate forces without damaging the component; ingenuous techniques to be adopted for the ease of clamping like quick acting screws, cam clamps, hydraulic clamps, etc.

(*iii*) Provision for easy loading and unloading of components

(*iv*) To hold the assembly together to withstand the cutting forces

(*v*) To design the location, size and material of the setting block to enable the cutter to be set in relation to the fixture and to be precise in relation to the component to be machined this is applicable only for milling fixtures

(*vi*) Provision for swarf removal

(*vii*) Fastening of fixture to the machine table or chuck or collet

(*viii*) Proper design of tenons at the bottom of the fixture so as to properly locate the fixture with respect to the machine table.

In addition to the above said points, the following aspects should also be taken into account in the design of jigs and fixtures:

(*i*) Consideration of sequence of operation given in the operation chart with particular reference to the operation to be performed

(*ii*) Study of the detailed drawing of the component critically, especially the dimensions which are provided with tolerances

(*iii*) Consideration of the manufacturing defects such as (*a*) shrinkages (*b*) blow holes, (*c*) inclusions as in the cast bodies of jigs, (*d*) distortions as in the case of welding and fabricating jig body or frame.

SUMMARY

In this chapter, a brief outline has been provided on the tooling involved in the manufacture of components to meet the requirements of accuracy and interchangeability with low cost of production. Various classes of engineering production, such as job, batch, flow and mass production are elaborated. The use of tooling, particularly in the "batch" and "flow" production models, is further explained. Definition of jigs and fixtures and their distinguishing characteristics are explained.

Various points on the design of jigs which need to be focused are: (*i*) location of components, (*ii*) clamping, (*iii*) guiding the tool in the case of jigs/setting the cutter in the case of milling, and (*iv*) loading/unloading of the components. Lastly, additional design features such as (*i*) focus on the tolerence dimensions of the component, (*ii*) study of the sequence of operations relating to the operation in question, and (*iii*) design for manufacturing are also listed.

REVIEW QUESTIONS

1. Draw a flow chart of the activities that follow while using a jig or a fixture when machining a component.
2. Distinguish between job, batch and mass production in manufacturing. Explain the functions of jigs/fixtures in each of the above types of production.
3. What will be the extent of increase in productivity by using a jig or a fixture? Explain with a specific example.

❑❑❑

Design of Locators

2.1 GENERAL PRINCIPLES OF DEGREES OF FREEDOM AND CONSTRAINTS

A parallelopiped shown in Fig. 2.1 has six degrees of freedom in space, namely, three translations along X–X, Y–Y and Z–Z axes and three rotational movements about the three axes. In order to provide constraints to the body, which has parallel and right-angled plane faces, six pegs are provided, three pegs in the X–Z plane, two pegs in the X–Y plane, and one peg in the Y–Z plane. These pegs provide the required constraints in six degrees of freedom. This is further explained below.

The three pegs provide a constraint in movement along the vertical direction parallel to O–Y. Similarly, the two pegs provided along the X–O–Y plane and the one provided along the Z–O–Y plane provide constraints in translation movements along the axes parallel to O–Z and O–X respectively.

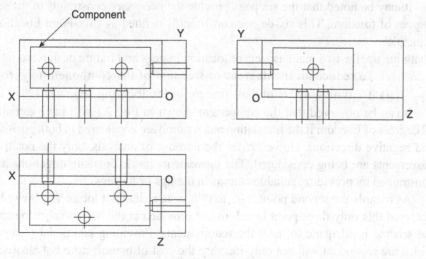

Fig. 2.1 Six Point Location Principle

Design of Jigs, Fixtures and Press Tools, First Edition. K. Venkataraman.
© K. Venkataraman 2015. Published by Athena Academic Ltd and John Wiley & Sons Ltd.

As regards the constraints for rotational degree about the axis parallel to the O–X axis, the three pegs in the X–O–Z plane and the two pegs in the X–O–Y plane provide the same. Similarly, the two pegs in the X–O–Y plane and the one located in the Z–O–Y plane provide the constraints in rotation about the axis parallel to O–Y. The three pegs at the horizontal plane and the one in the vertical plane provide the constraints in rotation about the axis parallel to the O–Z axis.

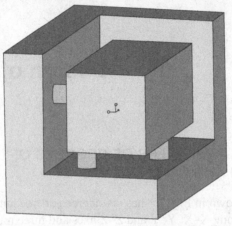

Fig. 2.2 Six Point Location Principle (3D View)

However, it may also be noted that the location of pegs, *viz.* the three pegs forming an isosceles triangle at the horizontal plane, as shown in Fig. 2.1, is one of the important factor in providing the desired constraint. This point is also applicable for the two pegs and the one peg located in the two vertical planes. These are provided midway in the height of the component, ensuring absolute constraint in rotation.

It may be noted that the six pegs provide the necessary constraints to the six degrees of freedom. This six-degree constraint is defined as "Six-Point Location Principle".

Summing up, the two main aspects of location in pegs and fixture design are:

    (*i*) To reduce all the degrees of freedom of the component to zero

    (*ii*) To avoid any redundant feature in the locating scheme.

It can be observed that the component shown in Fig. 2.1 will have actually 12 degrees of freedom if the translation and rotation are considered in both positive and negative directions. However, for the purpose of analysis, only the positive movements are being considered. The movements in the opposite directions are constrained by providing suitable clamps in the jigs or fixtures.

As regards the second point, *viz.,* to eliminate redundant locator(s), it can be observed that only three point locations are provided at the horizontal plane and the scheme is adequate to meet the requirements. Providing additional locators which are redundant will not only increase the cost of manufacture but also will increase the chance of errors in locating a component.

2.2 FOOLPROOFING

In the first chapter, it has been explained that the importance of using "jigs and fixtures" is to get increased productivity and reduce the overall cost. Therefore, when they are designed, attempts are made for the deployment of semi-skilled labour. This is again done for reduction in the production cost. Such being the case, the locators should be so selected that the component is loaded correctly with respect to the jig/ fixture, as well as in relation to the tool/cutter. This is more so in the case of unsymmetrical components, as shown in Fig. 2.3. In order to ensure that an unskilled or a semi-skilled worker can load the component correctly, four nos. of pins are introduced such that there is only one unique way of loading. In other words, even a fool can load the component correctly. Thus, the nomenclature "Foolproof" method is in vogue.

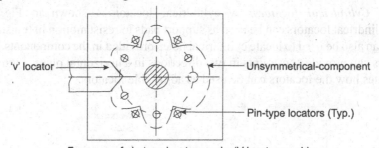

'v' locator ← → Unsymmetrical-component

Pin-type locators (Typ.)

Four nos. of pin-type locators and a 'V locator provide unique way of locating a component

Fig. 2.3 Foolproofing for an Unsymmetrical Component

2.3 OTHER PRINCIPLES IN THE DESIGN OF LOCATORS

Following are the general principles to be followed for the design of locators:

1. Sharp corners should be avoided in the locators. Hence, suitable chamfers or radius should be provided.
2. Locators should be of hardened material to withstand wear and tear of loading, clamping and cutting forces.
3. General class of locators is cylindrical, and therefore, close tolerances should be maintained in the specification as well as in the manufacturing process.
4. Locators should have relief groove at the interface where the diameters change. This enables the locators to sit squarely on the surface of the jig body/frame on which they are fitted, avoiding the burrs on the mating surfaces.
5. 'V' locators, both fixed and movable, should be used for locating cylindrical surfaces. This facilitates the centre of the cylindrical surface to be positioned exactly. 'V' locators are provided with chamfers along their vertical plane to provide a certain degree of clamping or hold-down effect and arrest the movement of the component. Figures 2.7 and 2.8 provide necessary illustrations for the 'V' locators.

6. Generally, peg or pin-type locators are short in height to resist bending forces. Height protruding above the peg surface may vary between 6 and 15 mm for general class of engineering applications. They are fitted to the jig / fixture frame by means of an interference fit. They could be screwed also. Height restriction is imposed so as to facilitate loading and unloading of components with ease.

7. Since the jigs and fixtures are used for producing the same components repeatedly, wear and tear of parts such as locators, bushes, clamps are common. Therefore, provision should be kept for replacement of such components.

2.4 VARIOUS TYPES OF LOCATORS

(*i*) *Cylindrical locators:* A cylindrical locator is shown in Fig. 2.4. The cylindrical locators can be used as support pads to resist motion in translation. They can also be used to locate cylindrical holes provided in the components. Such locators can provide constraints in two directions in a horizontal plane. Figure 2.5 illustrates how the locators can be used as adjustable locators.

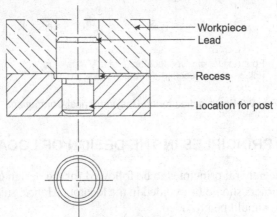

Fig. 2.4 Cylindrical Locator

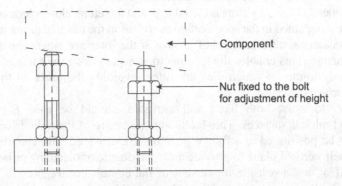

Fig. 2.5 Adjustable Locator

(ii) Long locators: Figure 2.6 explains the philosophy of using long locators. They are used in components having heights of 50 mm or more. The stem of the locator is reduced in diameter at the mid position to facilitate easy loadings and removal.

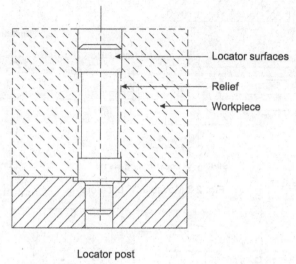

Locator surfaces

Relief

Workpiece

Locator post

Fig. 2.6 Long Locators with Relief at the Middle

(iii) 'V' locators, both fixed and sliding: As stated in the previous section, they are used to locate cylindrical objects. Figures 2.7, 2.8 and 2.9 illustrate such locators. They are downward chamfered at the locating faces to affect the clamping forces. Sliding 'V' locators can be cam-operated or can be simply screw-operated.

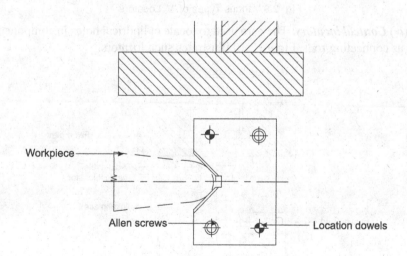

Workpiece

Allen screws

Location dowels

Fig. 2.7 Fixed 'V' Locators

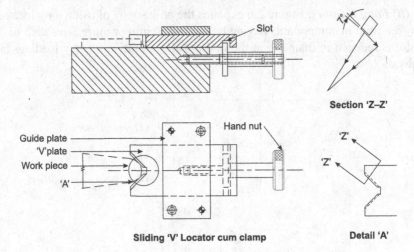

Section 'Z–Z'

Sliding 'V' Locator cum clamp

Detail 'A'

Fig. 2.8 Movable 'V' Locators

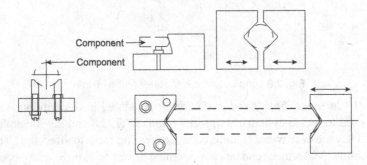

Fig. 2.9 Various Types of 'V' Locators

(iv) Conical locators: They are used to locate cylindrical holes in components such as connecting rods. Figure 2.10 illustrates such locators.

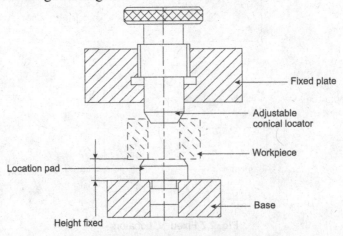

Fig. 2.10 Adjustable Conical Locators

(*v*) *Diamond pin Locators:* These are used in conjunction with principal cylindrical locators. Figure 2.11 illustrates the use of diamond pin locators and shows how the constraints in the two directions can be exercised in a connecting rod whose centre distance has tolerances.

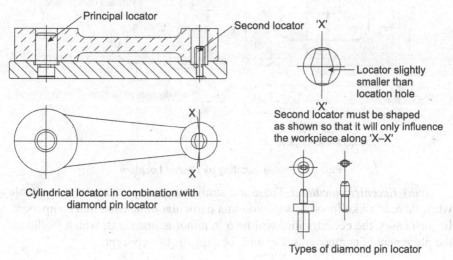

Fig. 2.11 Diamond Pin Locators

(*vi*) *Profile locating pins:* These are provided to suit the profile of the component, either square or curved. Figure 2.12 illustrates two examples of such locators.

(*vii*) *Nested locators:* In certain occasions, a casting could take a special profile, which needs to be machined, either drilled or milled. In such cases, the locators are so manufactured like a nest to suit the component profile, that the same sits in the predetermined groove (Nest). Figure. 2.13 explains the principle. In yet another occasion, the nested locators could cover a partial profile of the component, as shown in the same Fig. 2.13.

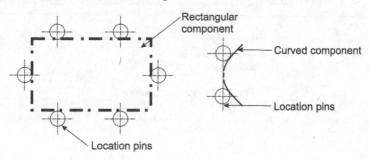

Fig. 2.12 Profile Locating Pins

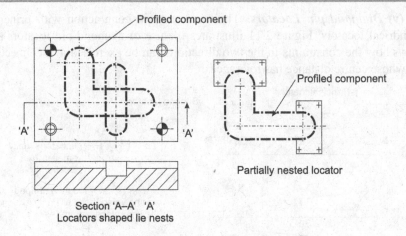

Section 'A–A'   'A'
Locators shaped lie nests

Partially nested locator

Fig. 2.13 Profile Locating by Nested Locators

(viii) *Eccentric locators:* These are similar to cam profile and are suitable when there is a likelihood of variation in a particular dimension in a component. In such cases, the eccentricities will help in minor adjustments, which facilitates the placement of components. Figure 2.14 explains this concept.

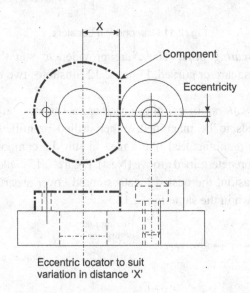

Eccentric locator to suit
variation in distance 'X'

Fig. 2.14 Eccentric Locator

SUMMARY

Function of locators and broad guidelines for their design are explained in this chapter. Concept of "Foolproofing" is elaborated. Principle of six-point location to exercise constraint in six degrees of freedom in a body is explained. Various types of locators such as (a) cylindrical, (b) adjustable (c) conical (d), 'V'-type (both fixed and sliding), (e) diamond pin and (f) nested locators are illustrated with figures.

REVIEW QUESTIONS

1. Why has the name "Diamond Pin Locator" evolved? Is it a principal locator or a secondary locator? Where is it generally used?

2. A flat surface of a machined component needs either four or three cylindrical locators at its bottom. Is it a correct statement? Why?

3. Why is the terminology "Foolproofing" used in locating unsymmetrical components?

4. Relief is provided in cylindrical locators. Substantiate your answer for short as well as long locators.

5. The sliding 'V' locators generally have vertically tapered edges at their locating faces. Why?

6. What are the specific advantages of conical locators?

7. Distinguish between nested locator and profile locating pins. What are their relative merits?

▢▢▢

Design of Clamps

3.1 PRINCIPLES OF CLAMPING

The basic functions of clamps are four-fold. They are as follows: (*a*) the workpiece must be held firmly even when the tools/cutters are in operation; (*b*) the clamping device should be quick acting as the loading and unloading time should be as quick as possible; (*c*) when subjected to excessive vibration or chatter, the clamps should be firm and should not loosen up; (*d*) the clamp should not damage the workpiece.

A tool designer defines clamping as, the holding of workpiece against the cutting forces; while the workpiece presses against the locating surfaces. There are innumerable types of clamping devices, which are designed or selected as per the requirements. If a large number of workpieces are involved, pneumatic or hydraulic clamps are also employed.

To design or to select a clamping device, the general guidelines to be followed are as follows:

- Simple clamping mechanisms should be adopted over complex ones, to save the cost of manufacturing and for ease of maintenance.
- Clamping parts, which are subjected to wear and tear, should be heat-treated so as to withstand cyclic operations. The material of the clamps should be so selected as to have properties like hardness, toughness and strength.
- Some frequently wearable parts of the clamps should be so designed as to be easily replaceable.
- Clamping force should be applied to a heavy part of the workpiece.
- Thrust of the cutting tool should be away from the clamp.
- Pressure pads should be employed wherever soft objects or hollow objects are to be clamped to avoid damage or distortion.

Design of Jigs, Fixtures and Press Tools, First Edition. K. Venkataraman.
© K. Venkataraman 2015. Published by Athena Academic Ltd and John Wiley & Sons Ltd.

3.2 CLASSIFICATION OF CLAMPS

There are different types o clamps. The design of the clamps, their selection, sizing, etc. depend on the component and the operation to be performed. Various mechanical types of clamps are illustrated in this chapter. In addition, hydraulic, pneumatic and electromechanical clamps are also used in applications where the rate of production needs to be comparatively high and the clamping forces need to be more rugged. The standard clamps that are generally used are discussed as follows: However, the clamps given are not the only alternatives. It is the tool designer's ingenuity to provide an efficient clamping system.

   1. *Clamps with heel pin:* These are of four different types. The heel acts as a fulcrum. The clamping force is applied at the middle through the screw and nut. The next is the point of contact with the workpiece, which holds the workpiece.

   - ***Solid clamp:*** This is used in drilling jigs and turning fixtures. They are common in many applications.
   - ***Clamp with heel pin:*** This has a stem like a heel and restricts rotation of clamps during clamping (Fig. 3.1).

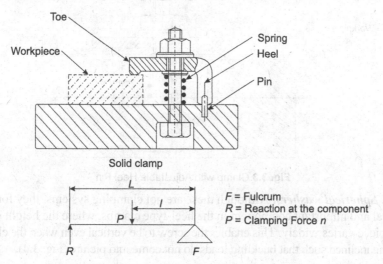

Fig. 3.1 Heel-Type Clamps

   - ***Slotted clamps with a heel pin:*** This is used when the rotation of the clamp is not needed as the clamp can be loosened and slid for the removal of components (Fig. 3.2).

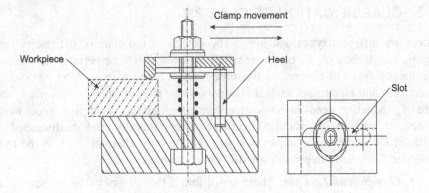

Fig. 3.2 Solid Clamps with Heel Pin and Slot for Quick Removal

- *Slotted clamps with an adjustable heel pin:* This is used when the component height is likely to vary and the adjustment of height of the clamp is imperative (Fig. 3.3).

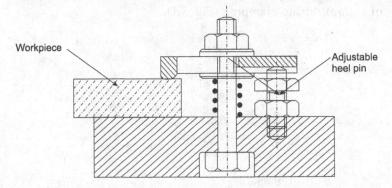

Fig. 3.3 Clamp with Adjustable Heel Pin

2. Spherical washers: Although these are not clamping systems, they form an integral and important component in the heel-type clamps, where the height of the workpiece varies widely. This enables the screw to be vertical even when the clamps become inclined such that buckling loads do not come into picture (Fig. 3.4).

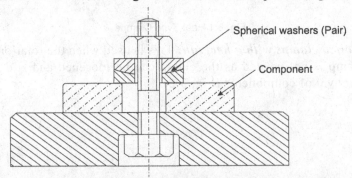

Fig. 3.4 Spherical Washers

3. *Two-point clamps:* These are used in clamping two components together, like in gang drilling operation (Fig. 3.5).

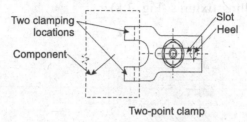

Two-point clamp

Fig. 3.5 Two-point Clamps for Distribution of Clamping Force

4. *Three-point clamps:* These are used for clamping hollow cylinders, for turning outside or slot milling inside keyways, and drilling oil-holes perpendicular to the axis (Fig. 3.6).

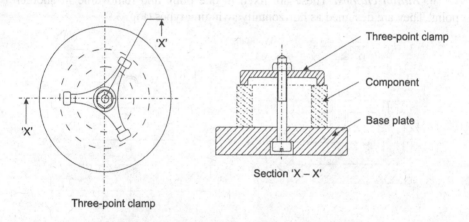

Three-point clamp

Section 'X – X'

Fig. 3.6 Three-point Clamp for Holding Circular Objects

5. *Latch-type clamps*

- **One-way clamps:** These are quick-acting, and are used for loading and removal of components. They are used in drilling jigs (Fig. 3.7).

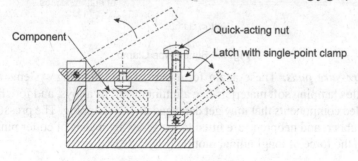

Fig. 3.7 Latch-type Clamp (one way)

- *Two-way clamps:* They are similar to the one-way clamps, except that the clamping forces are applied in two coordinates. They can be used in milling fixtures (Fig. 3.8).

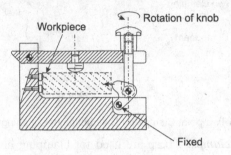

Fig. 3.8 Latch-type Clamp (two-way)

6. *Button clamps:* These are fixed in one point and removable in another point. They are designed as horizontally swinging types (Fig. 3.9).

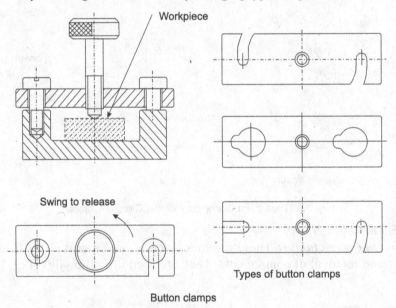

Fig. 3.9 Button-type Clamp

7. *Pressure pads:* These again form part of the clamping systems and are used while clamping soft materials like aluminum and its alloys, and for clamping thin-walled components that may get damaged due to clamping. The pressure pad, usually rubber, and neoprene are fitted to the screw by means of cotter pin, which transmits the force of longitudinal motion (Fig. 3.10).

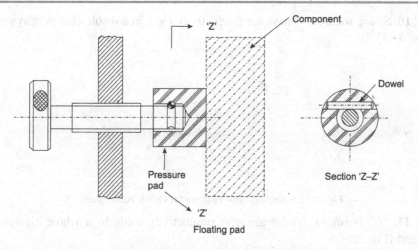

Fig. 3.10 Pressure Pad for Soft Material

8. *Wedge-type edge clamp:* These are used in cases like surface grinding, end milling, and surfatce facing of components. This will facilitate the exposure of the surface to be machined without interference with the tools (Fig. 3.11).

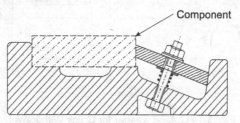

Fig. 3.11 Wedge-type Edge Clamp for Surface Machining/Grinding

9. *Equalising clamps:* These are used for clamping two components simultaneously, particularly for rough work like cutting (Fig. 3.12).

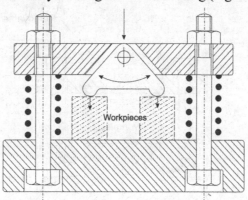

Fig. 3.12 Equalising clamp for Clamping Two Workpieces

10. *Swing washers:* These are productivity tools in a whole clamping system (Fig. 3.13).

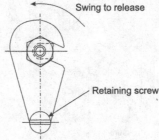

Fig. 3.13 Swinging-type Washer for Quick Withdrawal

11. *'C' Washers:* These are also productivity tools in a whole clamping system (Fig. 3.14).

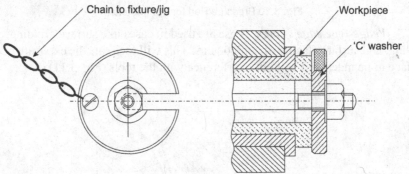

Fig. 3.14 C.-type washer for Quick Withdrawal

12. *Swinging bolts and removable-type clamps:* The bolts are designed to swing about a hinge and the clamps can be removed, allowing for unloading and loading of components (Fig. 3.15). They can be used for slotting, grinding and shaping fixtures.

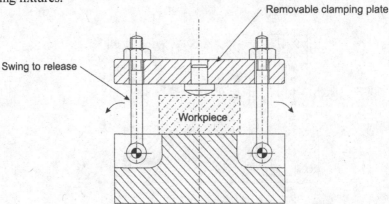

Fig. 3.15 Swinging Bolt-type Clamp

13. *Clamps for two components:* They are generally used in milling of keyways in shafts or drilling radial holes (Fig. 3.16).

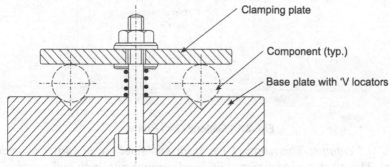

Fig. 3.16 Clamp for Two Components

14. *Cam clamps:* They utilise the profile of the cam for effectively applying the clamping force. Different types of can clamps are shown in Figs 3.17, 3.18 and 3.19.

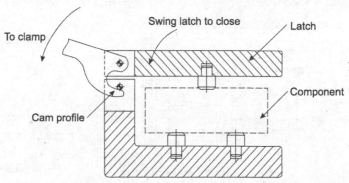

Fig. 3.17 Cam-operated Latch.type Clamp

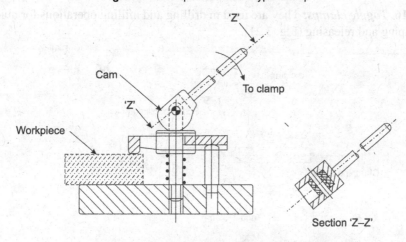

Fig. 3.18 Cam-operated Heel Type Clamp

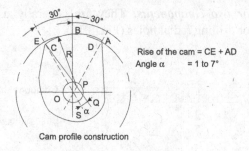

Rise of the cam = CE + AD
Angle α   = 1 to 7°

Cam profile construction

Fig. 3.19 General Cam Profile

15. *'V' clamps:* This concept has been explained in an earlier chapter on locators. These are used to clamp cylindrical components, both in the case of jigs and fixtures (Fig. 3.20).

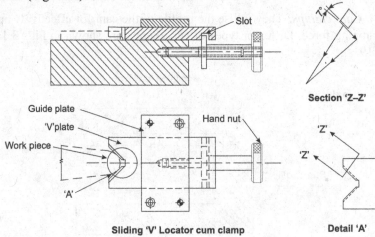

Sliding 'V' Locator cum clamp   Detail 'A'

Fig. 3.20 Sliding 'V' Clamp

16. *Toggle clamps:* They are used in drilling and milling operations for quick clamping and releasing (Fig. 3.21).

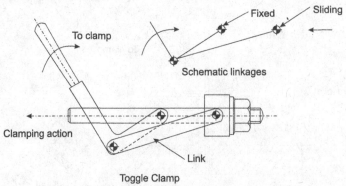

Toggle Clamp

Fig. 3.21 Toggle-operated Clamp for Short Operations

17. *Quick acting nut:* This forms a part of the clamping system to enhance its productivity (Fig. 3.22).

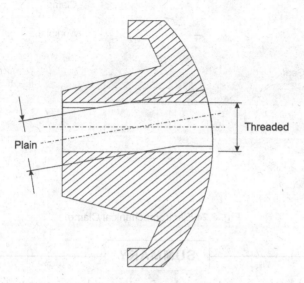

Plain

Threaded

Fig. 3.22 Quick-acting Nut

18. *Pneumatic clamps:* These are quick-acting, and are used for large-scale production (Fig. 3.23).

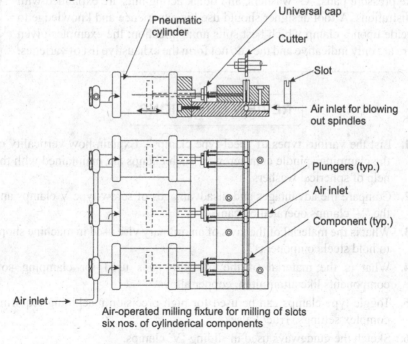

Pneumatic cylinder

Universal cap

Cutter

Slot

Air inlet for blowing out spindles

Plungers (typ.)

Air inlet

Component (typ.)

Air inlet →

Air-operated milling fixture for milling of slots six nos. of cylinderical components

Fig. 3.23 Air-operated Clamping Fixture

19. *Electromechanical clamps:* These are also used for large scale production and are illustrated in Fig. 3.24.

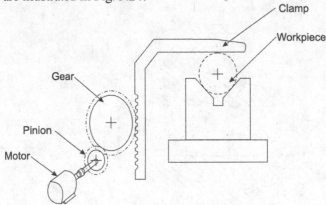

Fig. 3.24 Electromechanical Clamp

SUMMARY

In the introductory paragraphs, a brief outline of the points to be taken into account during the selection and designing of clamps has been explained. Further, the classification of general clamps and the subsystems of clamps like pressure pads, 'C' washers, and quick acting nuts are explained with illustrations. A tool designer should use his experience and knowledge to decide upon a clamp which best suits an application; the examples given here are only indicative and they do not form the exhaustive list of varieties.

REVIEW QUESTIONS

1. List the various types of "heel-type clamps". Explain how verticality of the clamping spindle cum screw of such clamps are maintained with the help of spherical washers.
2. Compare the advantages and disadvantages of screw-type V clamps and the 'V' clamps operated by cams.
3. What is the material of the jaws of an ordinary vise used in machine shops to hold steel components?
4. What is the material of the pressure pads used for clamping soft components like aluminium, copper alloys?
5. Toggle-type clamps can be used for high-precision machining requiring complex settings. True/ False?
6. Sketch the guideways used in sliding 'V' clamps.

7. Square threads are used in screw clamps when _____ forces are encountered.

8. 'V' blocks are used in clamping odd-shaped components. True/False?

9. A locator can be used for clamping as well as in one of the following cases:
   - Diamond pin locators
   - Conical locators
   - Nested locators
   - Fixed 'V' locators.

   Mark the correct answer.

10. Following type of clamping is preferred if large vibration and machine tool chatter are likely to be encountered.
    - Hydraulic clamping
    - Pneumatic clamping
    - Mechanical clamping

    Mark the correct answer.

11. How do you measure the clamping forces? Explain.

❑❑❑

Drilling Jigs

4.1 INTRODUCTION

In the manufacturing industry, making of holes, whether it is drilling, reaming, boring, punching or flame cutting, is one of the major activities. Therefore, tool designers and production engineers give utmost importance to proper location of the holes in the workpiece as well as on the machine, with the required tolerances and surface finish. It may be seen that the location and size of the hole are two important facets of mass production of components. This is achieved by a device known as drill jig, whose purpose is to locate and clamp the component firmly. The third important purpose is to guide the cutting tool, *viz.* drilling, boring, reaming and tapping. In the first chapter, the concept of interchangeability and accuracy were emphasized. These drill jigs enable unskilled workers to produce components that are accurate as well as interchangeable.

4.2 TYPES OF JIGS

The various types of jigs are given as follows:

1. *Plate jig:* Plate jigs are the simplest of all types of jigs. Here, a jig plate at the top containing drill bushes, is the main component of the jig. This is explained in Fig. 4.1.

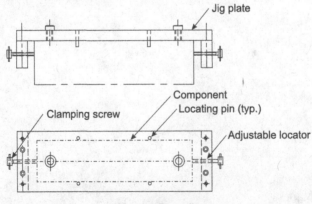

Fig. 4.1 Plate Jig

Design of Jigs, Fixtures and Press Tools, First Edition. K. Venkataraman.
© K. Venkataraman 2015. Published by Athena Academic Ltd and John Wiley & Sons Ltd.

2. **Solid-type jig:** Solid-type jigs are simple, yet they are rugged in design and are extensively used for components which are symmetrical and do not require much of foolproof locating devices. This is explained in Fig. 4.2.

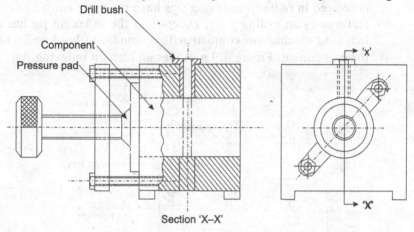

Drill bush

Component

Pressure pad

'x'

'X'

Section 'X–X'

Fig. 4.2 Solid Jig

3. **Box-type jig:** In box-type jig, the jig resembles a closed box with drill bushes in more than one side. This enables drilling or reaming operation to be performed in components like a flanged "T" joint or a four-way pipe joint having flanges in all the sides requiring drilling operation. This is illustrated in Fig. 4.3.

4. **Swinging latch type jig:** Latch-type jig has a latch operated through a hinge joint and the same rests on a square surface, so that the same is absolutely horizontal. Thus, the drill bushes fitted onto the latch plate will be absolutely vertical. Correct and incorrect latching techniques are explained in Fig. 4.4.

5. **Inclined jig:** Figure 4.5 is one of the examples for inclined jig. Whenever the component needs to be tilted about its axis so as to carry out the drilling operation of an inclined hole, such jigs are used. Such jigs are fabricated with precision tools off a metrology aboratory to ensure accuracy in the axis of tilt.

6. **Turnover jig:** Whenever component loading and unloading are done after the jig is turned over upside down, such a jig is known as turnover jig. Such designs are developed for reducing the cost of manufacturing of jigs for components requiring low levels of accuracies and repeatability. This jig is illustrated in Fig. 4.6.

7. **Pot jig:** The very name "Pot Jig" indicates that the jig body is in the form of a pot to accommodate cylindrical cup-like components requiring drilling operations. Figure 4.7 shows the cross -sectional view of a pot jig.

8. **Post jig:** Post Jig is designed to accommodate hollow cylindrical components. Figure 4.8 indicates a typical cross-sectional view of a post jig. The central post is the key component in the jig, particularly for firm gripping of the workpiece.

9. *Indexing jig:* An ingenious design concept in drilling jig is the indexing jig. This can perform more than one drilling operation in a component, in cases where the drill holes are spaced equally, if measured in radians. Indexing jigs have the same constructional features as an ordinary jig, except that the indexing jig has an indexing mechanism coupled with a spindle to hold and rotate the component. Figure 4.9 provides an idea on indexing jig.

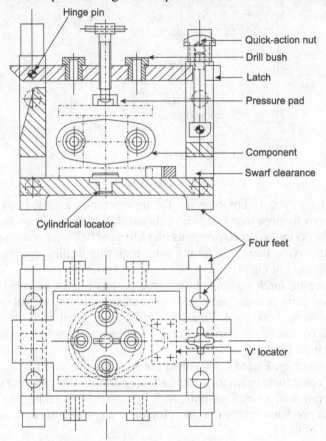

Fig. 4.3 Box-type jig

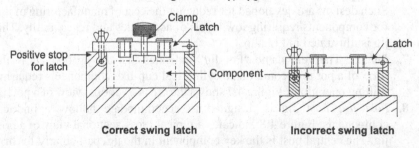

Fig. 4.4 Correct and Incorrect Latching Techniques

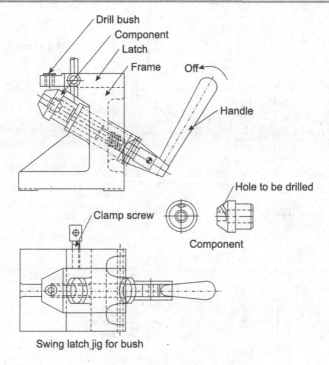

Fig. 4.5 Inclined Latch-type Jig

The basic principles followed during the construction of jigs are:

    (*i*)   Locating the component uniquely.

   (*ii*)   Clamping the component rigidly to withstand chatter and vibrations induced due to the cutting forces.

  (*iii*)   Rigid frame, either casted or fabricated, with a sound footing design. In certain specific cases where jigs are used in radial drilling machines, jig feet should be designed to facilitate clamping of the jig with the table of the machine tool.

  (*iv*)   Wear-resistant and precisely machined and locating the bushes to guide the tools.

   (*v*)   Easy loading and unloading facilities for the components.

  (*vi*)   Easy swarf disposal facility.

4.3 COMPONENTS OF JIG

A jig has basically the following components:

- Drill bushes which guide the tool to the component
- Fasteners which are quick-acting and robust
- Jig body or frame including the jig feet
- Indexing mechanisms

In the following paragraphs, the above items are explained in detail.

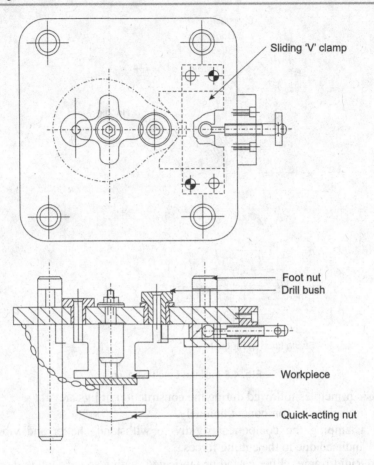

Sliding 'V' clamp

Foot nut
Drill bush

Workpiece

Quick-acting nut

Fig. 4.6 Turnover Jig

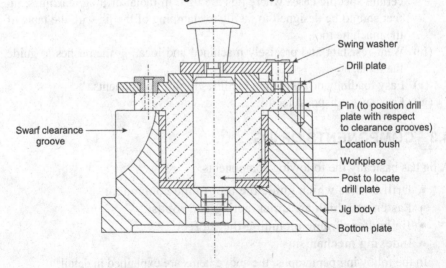

Swing washer

Drill plate

Pin (to position drill plate with respect to clearance grooves)

Location bush

Workpiece

Post to locate drill plate

Jig body

Bottom plate

Swarf clearance groove

Fig. 4.7 Pot Jig

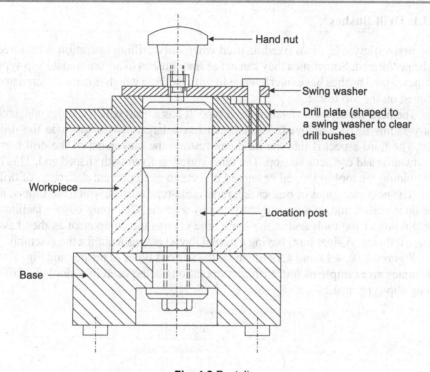

Fig. 4.8 Post Jig

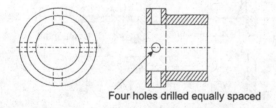

Fig. 4.9 Indexing Jig

4.3.1 Drill Bushes

The first variety, *viz.* plain fixed, is used when only drilling operation is required to be performed. Sometimes they can act as liner bushes to accommodate slip-type bushes. Fixed bushes have interference fit with the hole which is drilled, bored and reamed on the top frame.

The next variety is the headed fixed type. It has a head which enables adequate entry radius to be employed. Secondly, it has a larger length to guide the drill bits. The third aspect is that the head can restrain the movement of the drill head downwards and can act as a stop. The third variety is fixed with shaped end. This is for guiding the tool in shaped or curved objects so as to prevent deflection of drill bits. The next categories of bushes are, renewable-type bushes, which are slipped to the liner bushes and subsequently arrested by a screw. Clamping bushes facilitate the guiding of the tools and at the same time clamps the component as they have external thread. A liner bush having external thread acts as a nut for the assembly.

Figures 4.10, 4.11 and 4.12 illustrate the various types of bushes and Fig. 4.14 illustrates an example of four different operations being used in a single location using slip-type bushes.

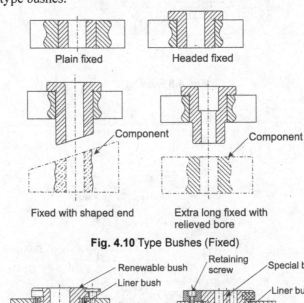

Fig. 4.10 Type Bushes (Fixed)

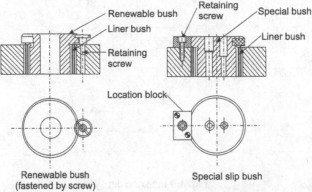

Fig. 4.11 Types of Bushes (Renewable)

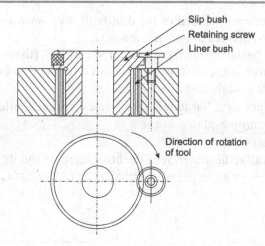

Fig. 4.12 Slip-type of Bush

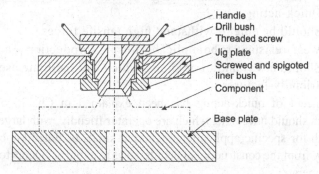

Fig. 4.13 Clamping Bush

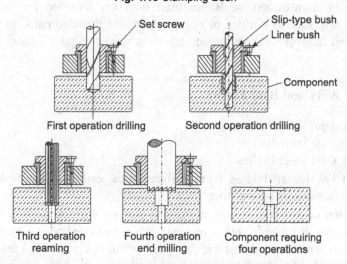

Fig. 4.14 Slip-type Bushes used for four Different Operations in the Same Location

Selection of correct material of the drill bush is an important aspect of the design. The following types of materias are used:

(*i*) Fixed bushes made from cemented carbides (these are expensive, but have longer life when compared to metallic bushes)

(*ii*) Nitrided, high-carbon steel bushes

(*iii*) Ordinary case hardened, like flame-hardened mild-steel bushes

(*iv*) Hard chrome-plated bushes

(*v*) Tool steel.

Generally, the material of the bush selected should be harder than the tool so as to prevent damages or indentations caused due to the tool.

4.3.2 Fasteners

The basic design concept for fasteners is given as follows:

(*i*) Quick-acting

(*ii*) Should be able to withstand high tensile forces

(*iii*) Wear-resistant, to be applicable for mass production of components

(*iv*) Should have handles or knobs, which will facilitate user/operator friendly features.

The concept of quick-acting has been explained in Ch. 3, *viz.* Clamps. The fasteners should have nuts, which are operator friendly, with large handles—large enough for specific applications. Knurled knobs or fluted knobs are to be located away from the constructional details, so as to avoid accidents to the fingers of the user/operator.

As regards to properties such as high tensile forces and wear resistance, low-alloy medium-carbon steels are used with the threaded portions being case hardened, such as nitrided or cyanide hardened threaded parts. In the case of clamping fasteners, coarse threads are to be used for quick action and easy operation.

4.3.3 Jig Body and Base Frame

Jig bodies are of the following types:

(*i*) Cast iron bodies

(*ii*) Cast steel bodies

(*iii*) Fabricated bodies from rolled plates, either using dowels and screws or by welding.

Cast iron or cast steel bodies are used in heavy constructional requirements, where vibration damping is a prime requisite. While cast iron bodies are brittle, cast steel exhibits superior toughness and strength. But this category of jigs and fixtures are becoming obsolete due to the advent and availability of a wide variety of rolled sections. Secondly, superior welding techniques with a wide selection of variables have led to the fabrication following this route. However, care is needed to be taken against distortions due to welding, by following proper welding procedure, such as zigzag welding or stitch welding, etc. Heat treatment subsequent to welding is also required to relieve thermal stresses induced due to welding.

Jig feet or base is also a part of the jig body. Jig feet can be classified as follows:
    (*i*) Cast or fabricated feet integral with the body
    (*ii*) Short column-like supports.

Cast or fabricated feet are clamped to the drill table. This variety of jig feet is extensively used. Whenever the jigs are to be turned over for loading and unloading the components, the second variety, *i.e.* short column-like supports are used. Both these types of illustrated in Fig. 4.15.

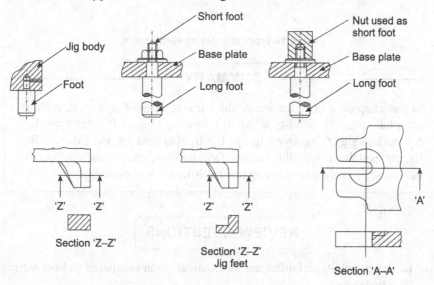

Fig. 4.15 Types of Jig Feet/Base

4.3.4 Indexing Mechanisms

Whenever more than one hole is to be drilled at equal spacing (*e.g.*, holes on a flange of a coupling), indexing plate with an indexing mechanism is used. In this case, the components are located over the indexing plate and are suitably clamped. The indexing plate is so designed as to enable rotation about an axis, either vertical or horizontal, as per the operational requirements. The indexing plate is provided with grooves at equal intervals, which again depends on the number of holes to be drilled at equal spacing. Indexing mechanism will have spring-actuated locator which will engage on the groove of the index plate, thus locking the mechanism. Figure 4.16 illustrates such mechanisms.

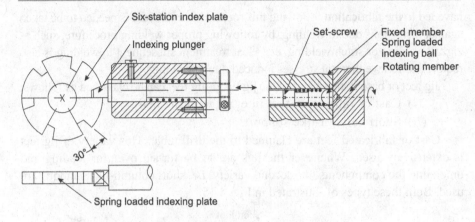

Fig. 4.16 Types of Indexing Mechanisms

SUMMARY

In this chapter, a brief outline of the various types of jigs used, such as (*a*) plate jig, (*b*) solid-type jig (*c*) box-type jig, (*d*) latch-type jig, (*e*) inclined jig (*f*) turnover jig, (*g*) pot jig, (*h*) post jig and (*h*) indexing jig, are explained with illustrations. Drill jig components, such as bushes, fasteners, jig body and indexing mechanisms are also explained.

REVIEW QUESTIONS

1. Renewable-type bushes are economical when compared to liner bushes. True/False?
2. What are the types of fastening or fixing the renewable bushes to the jig frame?
3. The type of fit at the outside diameter of a liner bush is:
   - Transition fit
   - Interference fit
   - Sliding/running Fit
   Mark the correct answer.
4. Box-type jig is used for drilling holes at angles in different faces of a component . True/False.
5. List the advantages and disadvantages of latch-type jig.
6. What will be the material used for the following components of a jig used in machining steel components:
   - Jig feet
   - Jig body
   - Clamps
   - Locators

7. In drilling operation, a clamp is used for one of the following reasons:
   - To resist the thrust of drilling
   - To counter the twist caused while drilling
   - To clamp the component from upward forces when the drilling tool breaks accidentally
   - To counter the twist caused while drilling as well as to clamp the component from upward forces when the drilling tool breaks accidentally
   - To resist the thrust of drilling as well as to counter the twist caused while drilling

   Mark the correct answer.

8. Machining accuracy of an indexing jig is considerably high. True/False?

9. Indexing pin is surface hardened in many cases. Why?

10. Sketch two commonly adopted index mechanisms.

5

Design of Milling Fixtures

5.1 SALIENT FEATURES OF MILLING FIXTURES

Milling is one of the important operations in production processes. The design of milling fixtures involves specific considerations to forces which cause the effects of vibrations or chatter. Rigidity of the total assembly, with particular reference to clamping, plays a vital role. It was explained earlier, that while clamps are designed as rugged, care needs to be taken against indentation or damage due to heavy clamping force. Vice-jaws are a very common clamping technique, with one of the jaws being fixed to the fixtures and the other one in the pair being of floating type. Milling fixtures have tenon strips at the bottom, which position the same in the milling table. The concept can be understood from Fig. 5.1. In addition, T-bolts are provided to fasten the fixture to the table. In addition to tenons and hold-down bolts, fixtures have setting blocks, which enable the cutter to be placed accurately in relation to the workpiece. Figure 5.2 illustrates a setting block. These are fastened by screws and dowel pins onto the fixture body. These are generally high -carbon steel ground and hardened. In earlier times, the body of the fixture used to be made up of cast iron to absorb vibration. But later on, cast steel or heavy fabricated sections started to replace cast iron due to the advantage of quicker methods of fabricating them. Thus, the components of a milling fixture are:

- Frame to absorb vibration and chatter
- Jaws or clamps to impart ruggedness in gripping
- Tenons and T-bolts for clamping the fixture to the milling table
- Setting block to identify and position the cutter accurately in relation to the workpiece.

Design of Jigs, Fixtures and Press Tools, First Edition. K. Venkataraman.
© K. Venkataraman 2015. Published by Athena Academic Ltd and John Wiley & Sons Ltd.

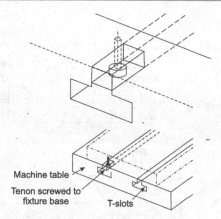

Fig. 5.1 Method of Fixing the Tenon to the Fixture Base

5.2   CLASSIFICATION OF MILLING FIXTURES

The following paragraphs explain the various types of milling fixtures and their applications. The corresponding figure numbers indicate the figures to be referred in each case.

5.2.1 Plain Milling Fixture

- *Special applications:* They are used for milling single components. The example shown in Fig. 5.3 has a locating pin in the centre and two heel-type clamps. Diamond Pin locator is provided to locate the auxiliary hole in the component. A pair of tenons are provided with T-bolts to clamp the fixture to the milling table. The said type has the setting block to align the cutter accurately with respect to the component to be machined.

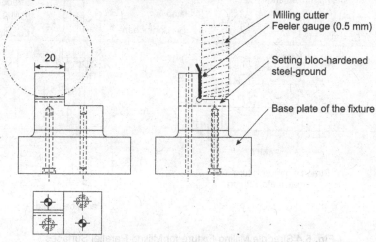

Fig. 5.2 Arrangement of a Setting Block

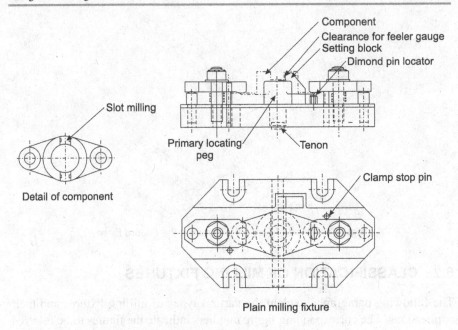

Fig. 5.3 Plain Milling Fixture for Machining Slot

5.2.2 Straddle Milling Fixture

- *Special applications:* This is similar to plain milling fixture except that here two milling cutters are used simultaneously to machine two parallel surfaces in a single component. In the type shown in Fig. 5.4, two locators are used with a cam-type clamp.

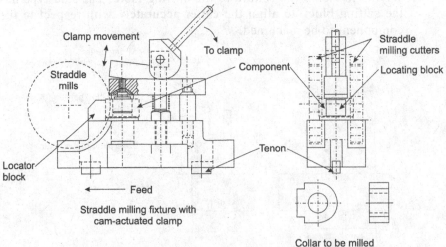

Fig. 5.4 Straddle Milling Fixture for Milling Parallel Surfaces

5.2.3 Gang Milling Fixture

- *Special applications:* Gang milling is an operation wherein two or more components can be machined simultaneously. Figure 1.2 shows gang milling of bolt heads. It may be noted that a single setting block is being used to set one of the cutters. The other milling cutters are positioned relative to the one positioned with respect to the setting block by means of collars placed on the arbour. Also, from Chapter 3, it can be observed that a clamp can be used for milling keyway slots in two shafts simultaneously. The number of components that can be machined together depends on the width of the machine table. If the machined part needs to be true, arbours need to be rigid and supported adequately. Alternatively, the number of components machined at a time can be minimized.

5.2.4 String Milling Fixture

- *Special applications:* String milling is an operation wherein more than one component gets machined one after the other. During this process, care needs to be taken to clamp all the components accurately in relation to the center line of the fixture, or to be more precise, in relation to the line of cutter movement and the feed. 'Equalizing clamps', as explained earlier in Chapter 3 are used to clamp pairs of components. Any other clamping system may not prove to be accurate, particularly when there is a likelihood of dimensional changes in two successive components.

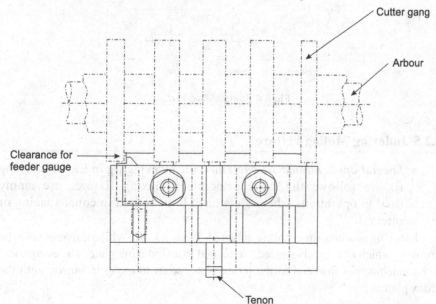

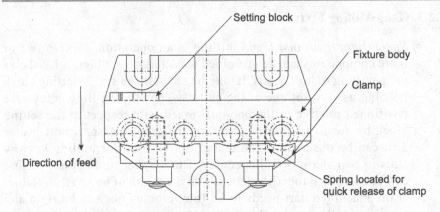

Fig. 5.5 Gang Milling Fixture for Simultaneously Milling Several Components

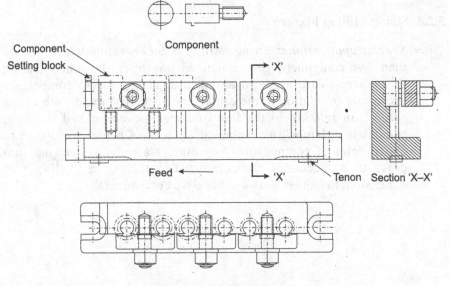

Fig 5.6 String Milling Fixture

5.2.5 Indexing Milling Fixture

- *Special applications:* As explained for indexing jig in Ch. 4, indexing fixture follows the same principle. Indexing fixtures are mainy used in operations where slots have to be milled in equal spacing or interval.

Indexing mechanism is spring actuated. It has a knob, which engages onto the grooves (which are equally spaced) machined to an indexing plate. The component to be machined is fastened to the index plate, so as to move in unison with the index plate.

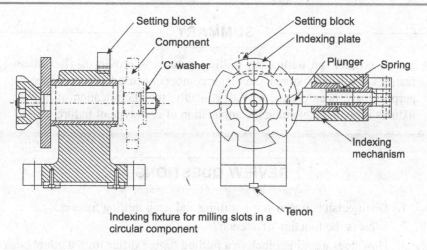

Indexing fixture for milling slots in a
circular component

Fig 5.7 Indexing Milling Fixture

5.2.6 Hydraulic Fixtures

● *Special applications:* Hydraulic fixtures are used in heavy milling operations requiring uniform clamping force. These fixtures are also used when mass production is adopted, when the operator is likely to get fatigued due to repeated physical operation of the clamps.

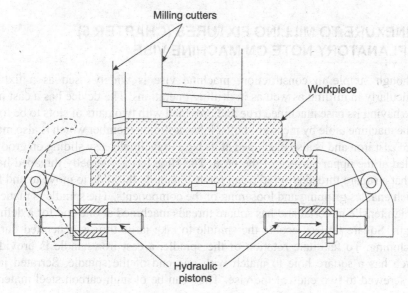

Fig. 5.8 Milling Fixture using Hydraulic Clamps

SUMMARY

In this chapter, an outline on milling fixtures is provided. The salient features of milling fixtures, the major components used and their specific purpose have been discussed. Subsequently, the classification of milling fixtures is given, explaining the application of each type of fixture.

REVIEW QUESTIONS

1. Distinguish between string milling and gang milling fixtures.
2. What is the function of a tenon?
3. How does a setting block in a milling fixture differ from a guide bush of a jig?
4. How is the base of a milling fixture designed? What are the factors to be taken into account for such a design?
5. List a few components which are machined in hydraulic and pneumatic fixtures. What are their relative merits?
6. 'V' blocks are widely used in milling fixtures. State the reasons.

ANNEXURE TO MILLING FIXTURES (CHAPTER 5) EXPLANATORY NOTE ON MACHINE VISE

Although simple in construction, machine vise is widely used as a fixture, particularly in milling as well as in drilling operations. The device has a cast iron base having its base machined true and provided with two pairs of slots to be fixed to the machine table by means of bolts. It has a sliding member which is also made up of cast iron and is also machined at its base to facilitate easy sliding on grooves milled at the upper surface of the base. The slide has a internally threaded hole, so that a square threaded spindle can rotate, enabling the slide to move to and fro, which enables gripping and loosening of the components. The spindle is made up of high tensile material and has square threads machined at one end for a definite length. Square threads enable the spindle to take large thrusts generated during machining. To facilitate rotation of the spindle, a cast iron handle is provided, which has a square hole to match with one end of the spindle. Serrated jaws are screwed to two ends of the vise. They can be of high-carbon steel material, hardened at their gripping faces.

Vises can be of swivelling type as well as tilting type, to provide for more number of degrees of freedom in machining.

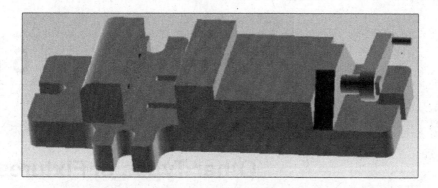

Fig. A5.1 Solid Model of a Vise Used in many Machining Operations such as Milling, Drilling, Boring and Shaping

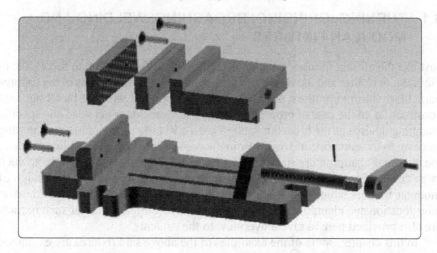

Fig. A5.2 Exploded View of an Ordinary Vise which could be Used as a Milling Fixture, Drilling Fixture and a Shaping Fixture

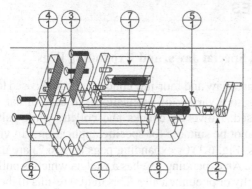

1	Base	1
2	Handle-1	1
3	Jan-2	1
4	Jan-1	1
5	Pin	1
6	Screw	4
7	Slide	1
8	Spindle	1

Fig. A5.3 Exploded Wire Frame Model of a Vise Providing the Bill of Materials

Other Types of Fixtures

6.1 TURNING, GRINDING, BROACHING, WELDING AND MODULAR FIXTURES

Just like the milling fixtures, which need to have specific attention to forces due to vibration or chatter and the rigidity of the equipment, the other types of the fixtures particularly turning fixtures, need the same amount of attention in the design and construction. In the case of broaching fixtures, the same does not need any specific clamping devices as the broaching action itself will provide the requisite clamping force on to the component. However, in the case of grinding and welding fixtures, location and clamping are important features, but they are not subjected to such large cutting forces. Now-a-days, developments are taking place in the design of modular fixtures which can be used in conjunction with computer algorithms to have location and clamping plans for polynomials. A brief outline of such fixtures are also provided here to give a overview to the students.

In this chapter, some of the examples of the above said fixtures are explained with relevant figures.

6.2 TURNING FIXTURES

6.2.1 Chuck Jaws, Collets, Special Jaws, and Expanding Pegs

Standard chuck jaws like the three-jaw and four-jaw chucks are widely used to hold work processes corresponding to a variety of applications. In case of bar work, spring-operated collets are used. However, chucks and collets, while fulfilling majority of applications, may not be suitable for specific applications. In view of these limitations, special jaws (Fig. 6.1) or expanding pegs (Fig. 6.2) are used to meet the specific requirements. An expanding peg has a wedge, which is pulled by a draw bar actuated mechanically or pneumatically. This in turn results in the three segments diametrically clamping the internal diameter of the component. But, in case of a wide variety of jobs requiring mass production, specially designed turning fixtures are used.

Design of Jigs, Fixtures and Press Tools, First Edition. K. Venkataraman.
© K. Venkataraman 2015. Published by Athena Academic Ltd and John Wiley & Sons Ltd.

6.2.2 Special Features of Turning Fixtures

In addition to general design features of fixtures, turning fixtures require the following points to be focused upon.

- The fixture along with the component should be perfectly balanced to avoid undesirable oscillations.
- The fixture should be free of any projections, so that interference with the tool post or cutting tool is avoided. In view of this specific criterion, sliding 'V' clamps are used. Wedge-type edge clamps are used so that they do not protrude outside. However, in certain cases where the component itself has a larger length, heel-type clamps can be adopted.

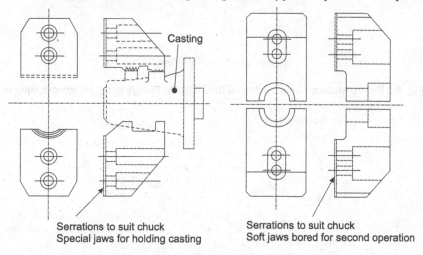

Serrations to suit chuck
Special jaws for holding casting

Serrations to suit chuck
Soft jaws bored for second operation

Fig. 6.1 Special Jaws to Suit the Turning Operations

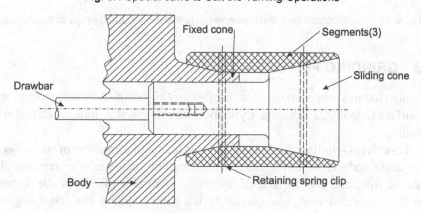

Fig. 6.2 Special Expanding Pegs

- Wherever fixtures are used in more than one machine, backplates have to be used. Backplates are just intermediate devices, positioned in-between the headstock and the fixture, and are either fastened to the flange of the spindle or screwed to the spindle directly, as shown in the Figs. 6.3 and 6.4, respectively.

Figure 6.5 shows a turning fixture for boring a bearing housing. It has a sliding 'V' clamp. A balancing weight is also shown in the figure. Such fixtures may be mounted on a backplate or faceplate.

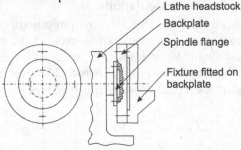

Lathe Backplates/Faceplates

Fig. 6.3 Fixture Mounted on a Backplate Bolted to the Flange of the Headstock Spindle

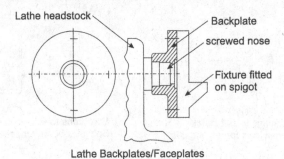

Lathe Backplates/Faceplates

Fig. 6.4 Fixture Mounted on a Backplate which is Screwed to the Spindle of Headstock

6.3   GRINDING FIXTURES

Grinding fixtures are generally classified for carrying out the following operations: 1. Surface Grinding, 2. External Cylindrical Grinding and 3. Internal Cylindrical Grinding.

1. Surface Grinding: fixtures are intended for holding a number of components on a plane surface, while the reciprocating motion of the grinder removes the material from the top surface of all the components placed in the fixture. The fixture in this case is a simple device with a sliding jaw, clamping the components onto the other sides of the fixture, which are rigid and fixed. The clamping could be effected either by screw movement or a cam or by pneumatic pressure.

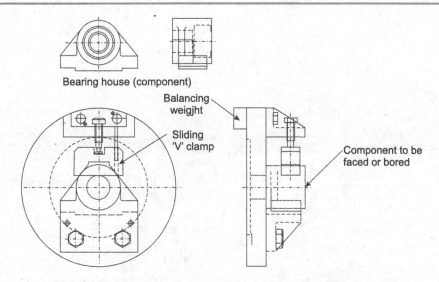

Fig. 6.5 Turning Fixture with Sliding 'V' Clamp for Boring Bearing Housing

2. External Cylindrical Grinding: In the external cylindrical grinding operation, a number of components are mounted on a mandrel, which rotates between centers. The driving peg transmits the rotation to the mandrel. A quick-acting 'C' washer and a nut effects clamping. This is explained in Fig. 6.6.

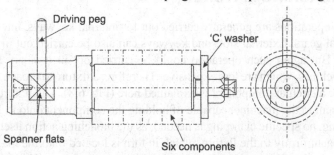

Fig. 6.6 Fixture for External Grinding Operation

In the case of *internal cylindrical grinding* operation, the fixture is mounted onto the headstock of the grinding machine. The component is mounted onto the fixture after locating with respect to a locator. Clamping is done by external heel-type clamps as indicated in Fig. 6.7.

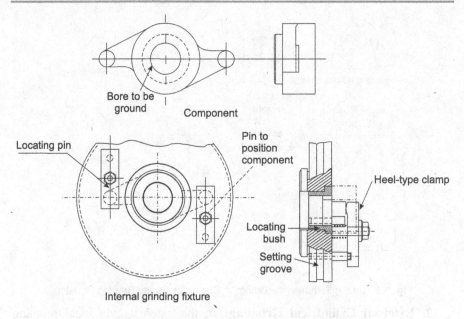

Bore to be ground

Component

Locating pin

Pin to position component

Heel-type clamp

Locating bush

Setting groove

Internal grinding fixture

Fig. 6.7 Fixture to Carry out Internal Cylindrical Grinding

6.4 BROACHING FIXTURES

Broaching operations are generally carried out for internal keyways, internal spline and internal gears. External slots and keyways can also be carried out in broaching machines. But again such operation could be carried out in ordinary slotting or milling machines, which are less expensive. Therefore, a fixture to broach an internal keyway for a circular component is explained here (Fig. 6.8). The fixture is simply an attachment to the machine, capable of holding the component onto it. In the case of broaching, no specific clamping is needed, as the broaching action itself will hold the component firmly to the fixture, which in turn is located onto the machine. The fixtures are made of high-strength material, which are tough as well.

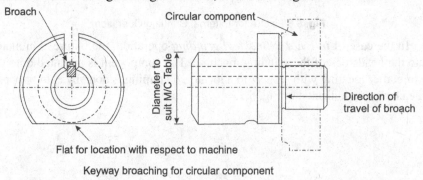

Broach

Circular component

Diameter to suit M/C Table

Direction of travel of broach

Flat for location with respect to machine

Keyway broaching for circular component

Fig. 6.8 Disposition of Internal Broaching Operation in a Circular component

6.5　WELDING FIXTURES

Welding fixtures are not required to be designed for heavy cutting and clamping forces. However, they are designed to resist distortions in welding operation. Locating pieces need to satisfy the geometrical criterion of the workpiece specified in the drawing. Two examples of welding fixtures are shown in Figs. 6.9 and 6.10.

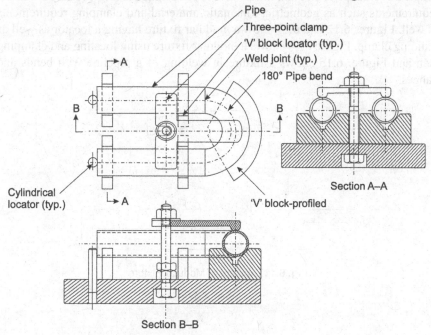

Fig. 6.9 Welding Fixture for Butt Welding of Pipes with 'U' Bend

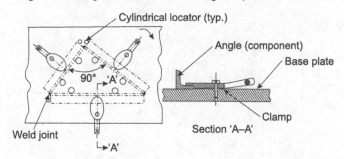

Fig. 6.10 Welding Fixture for a Frame using Locator Pins and Cam Clamps

6.6　MODULAR FIXTURES

Whenever the size and shape of the component or workpiece changes, dedicated fixtures may not be suitable, particularly in cases like fabrication work involving welding frames, and also in assembly requirements which require comparatively less clamping forces. In such occasions, modular fixtures can be used. Figure 6.11 shows the principle of working of modular fixture, wherein the circular "Fixels" can be located at random in circular holes forming a grid-like pattern. A polygon is located precisely in the figure, providing constraints in three degrees of freedom, namely two translation and one rotation, about a vertical axis. Although this is a simplest of examples, modular fixtures could be designed taking into account all the requirements such as geometric, kinematic, material and clamping requirements as well. Figures 6.12 and 6.13 show a modular fixture having a locator as well as a sliding clamp. Figure 6.14 shows a modular fixture using locating and clamping plan and Figure. 6.15 shows a fixture in welding of a pipeline with bends and flanges.

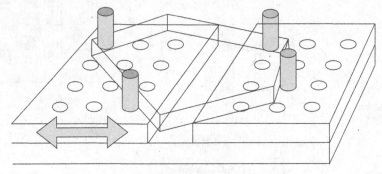

Fig. 6.11 Concept of Modular Fixture

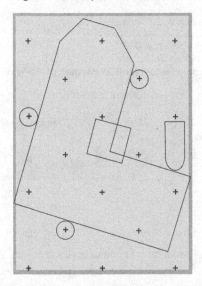

Fig. 6.12 Top Veiw of an L-shaped Component having locators and a Sliding Clamp

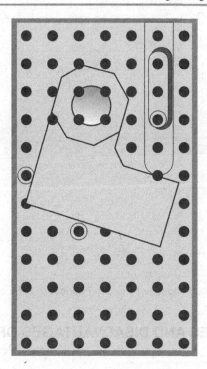

Fig. 6.13 Top View of an L-shaped Component having
Locators and a Sliding Clamp

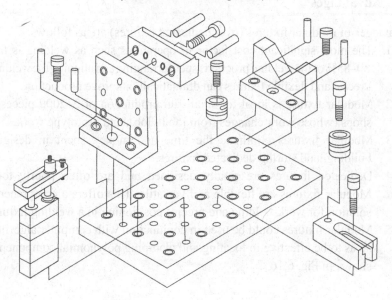

Fig. 6.14 Modular Fixture having Locators and Clamps

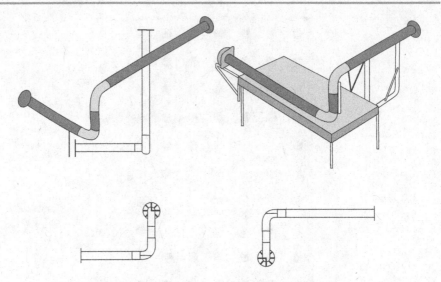

Fig. 6.15 Modular Fixture in Fabrication of Pipeline

6.7   ADVANTAGES AND DISADVANTAGES OF MODULAR FIXTURES

6.7.1 Advantages

Advantages of modular fixtures (over dedicated fixtures) are as follows:

1. The most significant cost factor in operations such as welding is fit-up. 80–85% of the welding process is spent arranging and aligning the weldment. Precise and flexible fixtures can dramatically improve production.

2. Modular fixture is ideal for small-run production of 2–2000 pieces, job shops (whose work changes from job to job), and prototype work.

3. Modular fixtures eliminate the time and money spent in designing, building, and storing dedicated fixtures.

4. Dedicated fixtures are rigidly designed and are often poorly tooled. Modular fixtures are flexible and versatile. This offers a comprehensive solution for work-holding and precise positioning in a welding fixture.

5. Modular fixtures could be used in conjunction with computer algorithms, so as to be effective in locating and clamping polynomial components as shown in Fig. 6.16.

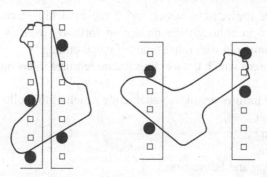

Fig. 6.16 Modular Fixtures being used for Locating Polynomial Components

6.7.2 Disadvantages

Disadvantages of modular fixtures over the dedicated fixtures are as follows:
1. Dedicated fixtures can be used for mass production of components, *viz.* for more than 10,000 components, without having to maintain the same.
2. Dedicated fixtures can be designed for much higher cutting forces, whereas modular fixtures have limitations in this aspect.

SUMMARY

In this chapter, five types of fixtures, namely, turning, grinding, broaching, welding and finally modular fixtures are explained. Each variety has a specific requirement for designing. Turning fixtures are needed to be designed to take care of rigid clamping as wells as errors arising due to imbalance. In the case of grinding fixtures, accuracy relating to circularity and eccentricity of components produced needs to be focused. Design of broaching fixtures is different from other classes of fixtures, as they are just work-holding devices having adequate strength and toughness to withstand cracks and fractures arising out of fatigue. Welding fixtures do not require high tolerance and dimensional accuracies, as in the case of milling or turning fixtures. Nevertheless, they are designed to produce components/ structures/ frames which meet the overall geometric requirements of the drawing and specification.

REVIEW QUESTIONS

1. What are the design factors on which the turning fixture differ from that of the milling fixture? What is the most important factor to be considered in turning or boring fixtures?

2. What are the reasons which make the broaching fixture simple in its construction although the broaching force by itself is quite enormous when compared with other types of metal cutting?

3. Major forces which the welding fixture requires to encounter is ....

4. Modular fixtures are more suitable for which of the following operations:
   - Milling
   - Turning
   - Drilling
   - Welding and fabrication
   - Assembly
   - Inspection.

5. State how the application of modular fixtures are most effective with the use of computer algorithms.

6. Design a simple milling fixture to mill a slot for the component as shown in figure below:

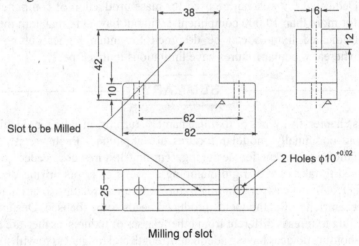

Milling of slot

ANNEXURE I TO OTHER TYPES OF FIXTURES
EXPLANATORY NOTE ON STEADY REST USED IN CENTRE LATHE

Although Steady Rests used in centre lathes cannot be classified under "Turning Fixtures", they also form part of the family of work-holding devices used to provide support to slender components while they are being turned. Figures A6.1 and A6.2 show respectively the solid model and the exploded view of the steady rests normally used in lathes. They have a cast iron base with a provision to slide on the lathe bed. This is the bottom half of this device. Top half is also made up of cast iron and the same can be bolted and fastened to the bottom half. A set of three screws are provided (equally spaced along the circumference of the circular face of the steady rest) to support the workpiece which is long and is being held between the live and dead centres.

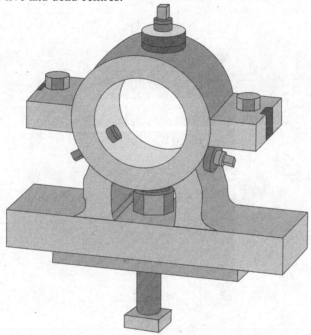

Fig. A.6.1 Solid Model of the Steady Rest Used in Centre Lathe

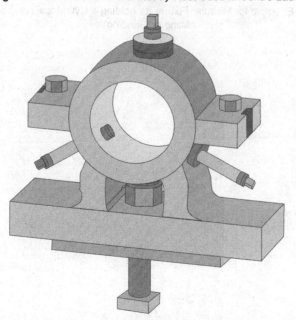

Fig. A.6.2 Exploded View of the Steady Rest

ANNEXURE II
OTHER TYPES OF FIXTURES

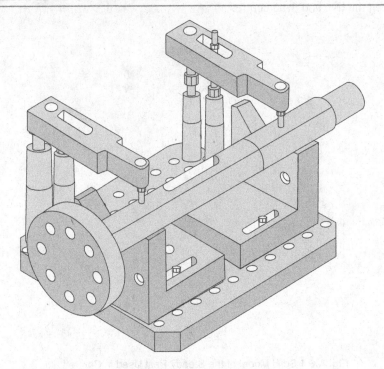

Fig. A.6.3 First Example for Modular Fixture for holding a Cylindrical component (Key Way Milling or Grinding)

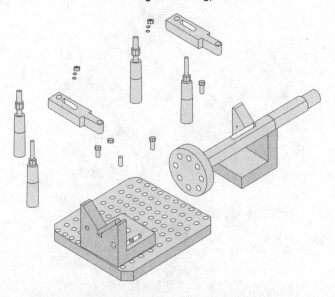

Fig. A.6.4 Exploded View of the Modular Fixture

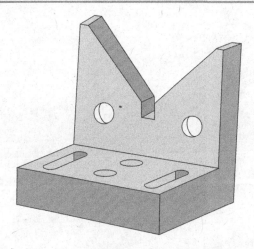

Fig. A.6.5 Individual "V" Block

Fig. A.6.6 Individual Component of the modular Fixture

Fig. A.6.7 Individual Component of the modular Fixture

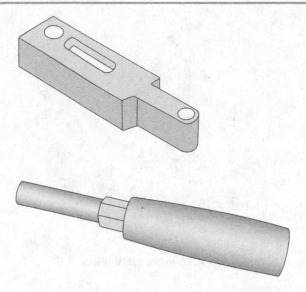

Fig. A.6.8 Individual Components of the modular Fixture

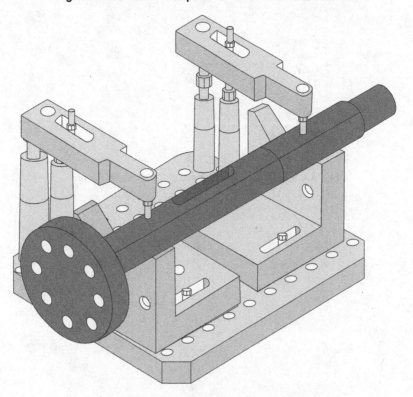

Fig. A.6.9 Assembled View of the modular fixture

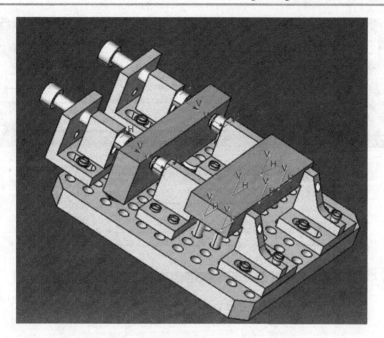

Fig. A.6.10 Second Example of Modular Fixture(For Surface Grinding and Preparation of a Component)

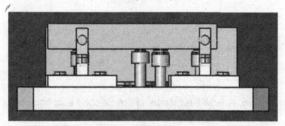

Fig. A.6.11 End View of the Second Example

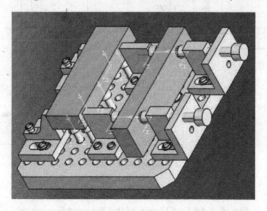

Fig. A.6.12 Perspective View of the Second Example

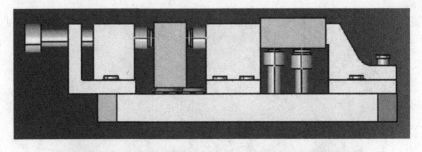

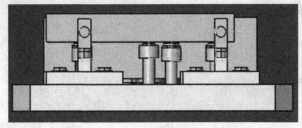

Fig. A.6.13 End Views of the Second Example

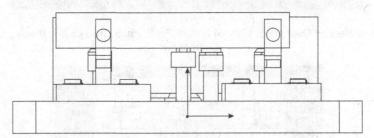

Fig. A.6.14 End Views of the Second Example

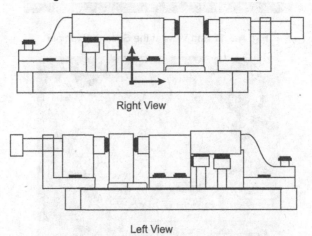

Right View

Left View

Fig. A.6.15 End Views of the Second Example

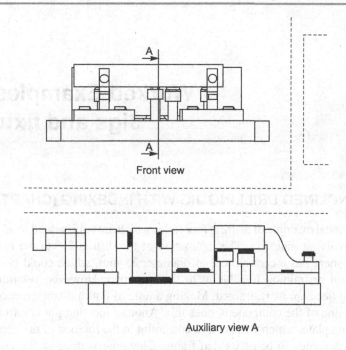

Front view

Auxiliary view A

Fig. A.6.16

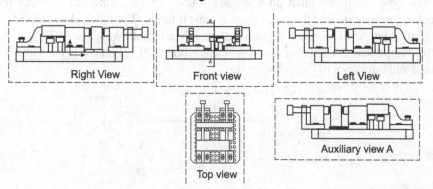

Right View

Front view

Left View

Top view

Auxiliary view A

Fig. A.6.17 All the Views of the Modular Fixture(Second Example)

Worked Examples for Jigs and fixtures

WE.1 INCLINED DRILLING JIG WITH INDEXING (CHAPTER 4)

This is a typical example of drilling four holes which are equi-spaced in an inclined fashion involving indexing. The inclination of the drill is 20° to the horizontal. The component has a central hole of diameter 25 mm, which could be used for locating and restraining translation in the two axes. However, rotation of the component needs to be restrained. Making a hole of 6 mm diameter, eccentric to the centre line of the component, does this. Another locating pin is introduced at the indexing plate, which will act as a constraint in the rotation of the component. The jig is designed to be of welded frame. Clamping is done at one end with a screw-type clamp mounted on a welded frame. The indexing pin provides the locking arrangement of the index plate against rotation. The index plate is fixed on a pin, which revolves about a gunmetal bush bearing allowing for 360° rotation. The component is shown in Fig. WE.1.1. Figures WE.1.2 and WE.1.3 show the front and top views of the indexing jig.

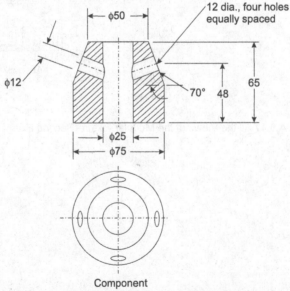

Component

Fig. WE.1.1 Component for Inclined Drilling Requiring Indexing

Design of Jigs, Fixtures and Press Tools, First Edition. K. Venkataraman.
© K. Venkataraman 2015. Published by Athena Academic Ltd and John Wiley & Sons Ltd.

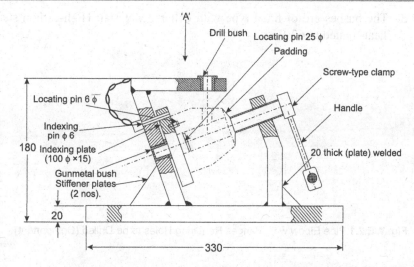

Fig. WE.1.2 Indexing Jig for Inclined Drilling (Front View)

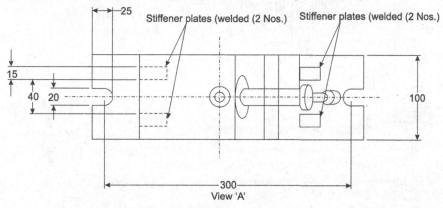

Fig. WE.1.3 Indexing Jig for Inclined Drilling (Top View)

WE.2 BOX JIG (CHAPTER 4)

Problem: Design a box jig for a pipe elbow having flanges at the two ends which are orthogonal. The jig is intended for drilling holes at the two faces of the flanges, four in one side and two in the other.

Solution: The drilling jig designed has the following salient features:

1. A nested locator at the box jig to locate as well as to support the curved portion of the elbow. Material: Wear-resistant cast steel.
2. The jig has two conical locators at the two orthogonal sides to locate the bore as well as to clamp the component. Material: High-tensile steel.
3. It has a latch which rests on the machined surface of the box jig which has cut-out recess machined to the required accuracy and tolerance.
4. The body of the jig is of cast steel material with proper footing on two of its sides to facilitate transfer of downward thrust.

5. The bushes are of fixed type with collars. Material: High-carbon steel, heat-treated to 60-Rc.

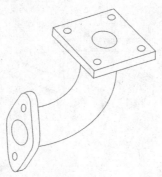

Fig. WE.2.1 Pipe Elbow with Flanges Requiring Holes to be Drilled (Component)

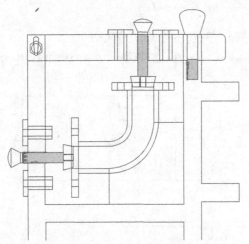

Fig. WE.2.2 Front View (Wire Frame Model) of the Box Jig Designed for Drilling Flanges of a Pipeline Elbow

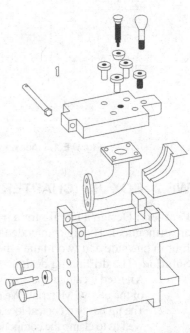

Fig. WE.2.4 Exploded View of the Box Jig Designed for Drilling holes in the Two Flanges

Fig. WE.2.3 Solid Model of the Box Jig

WE.3 INDEXING MILLING FIXTURE (CHAPTER 5)

The chosen component has two sets of slots to be milled at an angular difference of 90°. Hence, an indexing milling fixture is designed to meet the requirements. The fixture has an indexing plate which revolves about its spindle. Rotation of the index plate spindle is designed to take place over a cylindrical bush made of wear-resistant material like brass. Two sets of sliding 'V' clamps, consisting of two nos., each, are provided to clamp the component as well the index plate during the milling operations. Indexing mechanism consisting of indexing pin, cylinder, spring and a knob, is provided to carry out the indexing operation. Setting block made of hardened steel and a pair of tenons are also provided. Rotation of the index plate will be 90° to-and-fro and will not be for complete revolution.

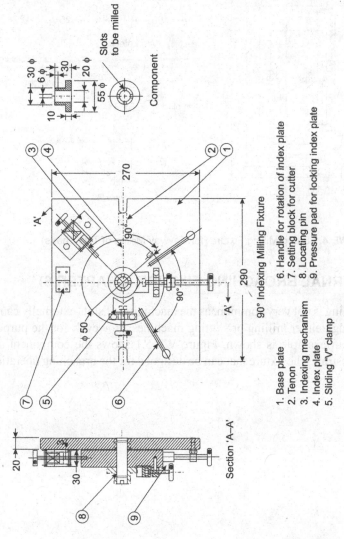

90° Indexing Milling Fixture

1. Base plate
2. Tenon
3. Indexing mechanism
4. Index plate
5. Sliding "v" clamp

6. Handle for rotation of index plate
7. Setting block for cutter
8. Locating pin
9. Pressure pad for locking index plate

Fig. WE.3.1 Indexing Milling Fixture

WE.4 STRING MILLING FIXTURE (CHAPTER 5)

This example is a design of a string milling fixture for milling keyways in two shafts. The milling operation is assumed to take place in a vertical milling machine with a vertically positioned cutter. The fixture has an equalizing clamp to hold the two shafts together. The shafts are located by means of locating pins positioned at the end. These locating pins not only assist in correct location, but also resist the thrust developed due to the cutting operation. Setting block and tenons are provided to position the cutter in relation to the component and to locate the fixture accurately in relation to the table/feed.

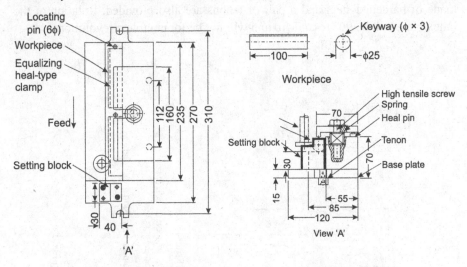

Fig. WE.4.1 String Milling Fixture (for Keyway Milling of Two Shafts)

WE.5 EXTERNAL BROACHING FIXTURE (CHAPTER 6)

External broaching is not very common, as the machining of slots externally can be performed through either milling or slotting machines. However, for the purpose of illustration, an example is shown. Figure WE.5.1 shows the component and Fig. WE.5.2 illustrates the fixture that can be designed for the broaching operation.

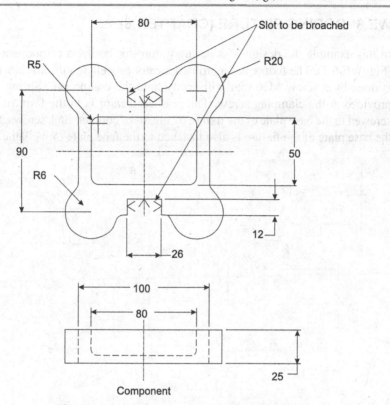

Fig. WE.5.1 Component for External Broaching

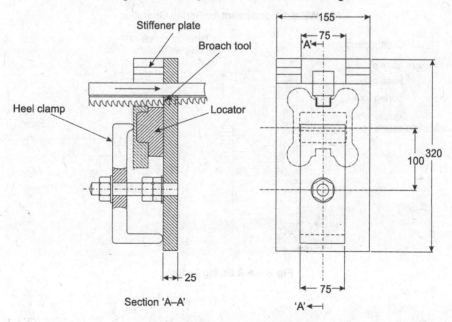

Fig. WE.5.2 External Broaching Fixture

WE.6 BORING FIXTURE (CHAPTER 6)

In this example the design of a boring fixture for boring a component is shown (Fig. WE.6.1). The fixture has a profiled locator as well as a locator pin. Clamping is done by a screw M16 that will firmly hold the component. Spring washer is provided at the clamping screw. The profiled locator is in the form of a bracket screwed to the base plate of the fixture by means of counter-sunk screws. Similarly, the base plate of the fixture is also fastened to the face plate of the lathe.

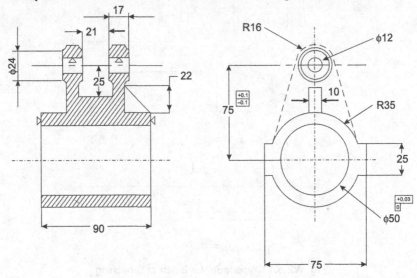

Fig. WE.6.1 Component for Boring Operation

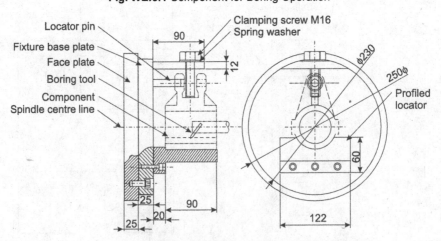

Fig WE.6.2 Boring Fixture

APPENDIX

Metal Cutting Tools

A.1 INTRODUCTION

The subject of metal cutting tools is quite vast. However, to supplement the study on jigs and fixtures, a brief outline on the various cutting tools which have relevance to the study on jigs and fixtures, is presented in this appendix. Geometrical details of the tools employed in cutting operations and the forces, such as thrust and torque, which are associated with such operations, are enumerated with sketches. Tools such as gear-hobbing and gear-shaping tools are not discussed, as they are not associated with general-purpose operations connected with jigs and fixtures.

A.2 SINGLE-POINT CUTTING TOOLS USED IN TURNING AND BORING FIXTURES

Tools are generally classified as single-point and multi-point cutting tools. Single-point cutting tools could be classified as solid, butt-welded/tipped tools and throwaway tool tips. Solid tools are produced from a solid piece of shank, as in the case of some of the turning and boring tools. They are made from forged high-carbon low-alloy steel, subsequently ground and heat-treated to the required hardness to withstand wear-resistance and to have hot-hardness properties. Butt-welded and tipped tools follow the same philosophy of either butt-welding an alloy tool material to the shank of the tool body or brazing the special cutting material to the point end of the tool. Throwaway tools use inserts screwed to the shank. These inserts are made of cemented carbides and have the advantage such as elimination of tool grinding time. A simple turning tool has a number of variables in its geometry such as:

(*i*)	Back Rake Angle	(*ii*)	Side Rake Angle
(*iii*)	End Relief Angle	(*iv*)	Side Relief Angle
(*v*)	End Cutting Edge Angle	(*vi*)	Side Cutting Edge Angle
(*vii*)	Nose Radius		

The above information has a bearing on tool life, quality and rate of production. The information is also called 'tool signature'. Barring rake angles, which could be either positive or negative, the other angles are measured positive and are generally less than 10° for carbide single-point tools. In the case of alloy cast steel tools, rake angles can go even up to 20° for machining materials such as stainless steel. The geometry of lathe tool is indicated in Fig. A.1.

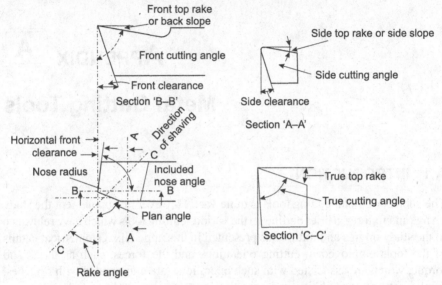

Fig. A.1 Nomenclature of Single-point Cutting Tool

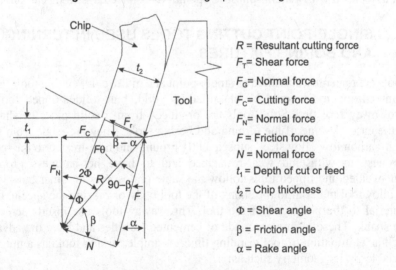

Fig. A.2 Forces Acting in Orthogonal Cutting

Cutting operation in a single-point cutting tool and the forces involved are modelled as shown in Fig. A.2. Figure A.2 represents an orthogonal cutting system (representing two-dimensional cutting operation) in which the cutting edge is perpendicular to the direction of motion relative to the work piece and is wider than the chip. F_C and F_N are cutting and normal forces, respectively.

These forces result in a resultant force R which could be measured by a dynamometer or through a transducer. The other forces which can be obtained from the resultant force are indicated in the figure along with the angles such as rake angle, shear angle and the friction angle.

The above geometrical analysis may be useful for carrying out the study of various forces in relation to the tool geometry, material to be turned, cutting tool material, and the cutting parameters such as speed and feed. However, in reality, three components of resultant force will be involved, namely, tangential force F_T, radial force F_R, and the feed force F_F. The same are shown in Fig. A.3.

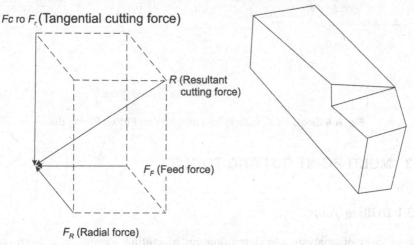

Three-dimensional Components of cutting forces

Fig. A.3 Resultant Force in Three-dimensional Cutting Operation

The tangential force F_T (in Newtons) can be calculated for a cutting tool material of high-speed steel by a rough formula :
$$F_T = C f^{n_1} d^{n_2}$$
where   C is the constant of proportionality, ranging between 1000 and 2000

f is the feed rate (in mm per revolution)

d is the depth of cut (in mm)

n_1 and n_2 are slopes of the plot between force *versus* feed and force *versus* depth of cut.

Here, n_1 varies between 0.43 and 0.84 and n_2 varies between 0.77 and 1.21.

(Variations in constant of proportionality and in the values of n_1 and n_2 depend on the type of material and its hardness, like hardened, cold finished, annealed, etc.)

Thus, for a feed rate of 0.5 mm per revolution and a depth of cut of 0.5 mm, a force of 700 N may be encountered in a tangential direction at the point of cutting. In addition, moments due to such forces may also be considered depending on the component dimensions. This will provide an idea to the designer of a turning or a boring fixture about the amount of clamping forces that may be needed to counter such forces. In addition, the fixture should be so designed as to dampen any machine tool chatter that may arise due to the sudden increase in the depth of cut, non-homogeneous material, tool wear, etc.

Figure A.4 shows a piloted boring bar fitted with single-point cutting tools. As explained earlier, the boring tool could be from solid shank, particularly in the case of short objects.

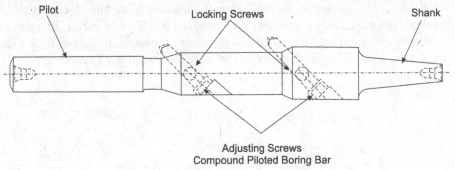

Compound Piloted Boring Bar

Fig. A.4 Single-Point Cutting Tool mounted on Piloted Boring Bar

A.3 MULTI-POINT CUTTING TOOLS

A.3.1 Drilling Tools

The process of analysis and determination of cutting forces for a multi-point cutting tool is more complex due to the variation in chip thickness and also due to the change in orientation of the cutting tool with respect to the component. In many cases, the forces are determined by determining the forces for a single-point tool and summing up the forces for all the tools involved. Initially, the forces encountered in a drilling machine are discussed.

Commonest of all the drilling tools is the twist drill, which has a chisel edge as the primary cutting edge at its bottom. Chisel edge corner and the outer corner of the drill bit also act as cutting edges, while the drill bit rotates about its vertical axis. The bottom portion of the drill bit is tapered, generally at an angle of 118° to allow for penetration of the tool while the cutting operation takes place. This angle is called the point angle and can vary between 90° and 140° for metal cutting. The flutes or the recess part along the length of the drill bit are spiraled. The helix angle may vary between 10° and 45° and is generally ground between 22° and 27°. The edges of the flute have rake angle similar to the single-point cutting tools. Figure A.5 shows the geometrical details of a drill bit. The drill bit diameters are generally back-tapered. Material for twist drills is generally HSS to withstand wear at high temperature.

In addition to twist drills, which are used for general-purpose machining, micro drills have varieties such as (a) single-fluted spiral, (b) straight double fluted, (c) two-fluted spiral, (d) pivoted and (e) flat drills. These do not use the drilling jigs because high-speed drilling have sensitive drill holders and frequent withdrawals are needed to remove the swarf.

In case of the design of a drill jig, an approximate estimation of forces is needed to evaluate the component sizes, particularly in clamping. Forces associated with drilling are explained in Fig. A.6. It can be seen that the drilling torque is a major factor to be restrained as the thrust is directed downwards and it generates an equal and opposite reaction at the support points of the drill table. Theoretically, a clamping force need not be required to counter the thrust. However, in case of a sudden drill bit breakage during the operation, restraining forces are needed to clamp down the component against the upward pull.

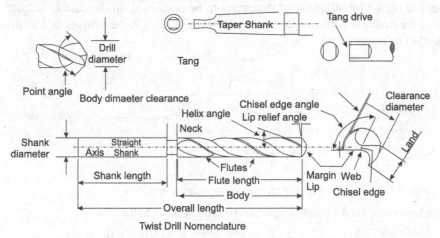

Twist Drill Nomenclature

Fig. A.5 Gometrical Details of Drill Bit

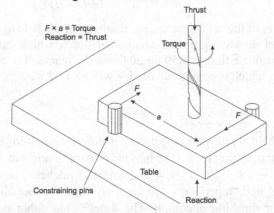

Fig. A.6 Forces Associated with Drilling in a Drill Fixture

The following paragraph explains the formula used to calculate the forces in drilling when the normal ratio of 0.18 is maintained between the width of the chisel edge and the diameter of the drill bit.

A rule of thumb formula is given as follows:

Torque in drilling M (in N mm) : $500\,f^{0.8}d^{1.8}$

Thrust in drilling T (in N) : $1740\,f^{0.8}d^{0.8} + d^2$

where f = feed per revolution (in mm)

d = diameter of the drill (in mm)

A.3.2 Reamers

Reamers are multi-point tools used for enlarging the diameter of the pilot hole to arrive at an accurate diameter with superior finish. Thus, the torque and thrust depend on the initial and final diameters to be machined. In the case of reamers used for multiple cutting operation as shown in Fig. A.7, the torques and thrusts need to be summed up.

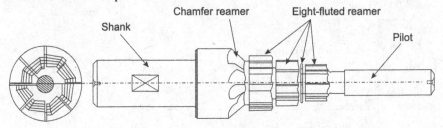

Fig. A.7 Multi-stepped Reamer with Pilot

The torque involved in reaming operation will be(N-mm): $\mathbf{580}\,kf^{0.8}\,d^{1.8}\,\dfrac{\{1-(d_1/d)^2\}}{\{1+(d_1/d)^{0.2}\}}$

The thrust involved will be (N) : $\mathbf{400}\,kf^{0.8}d^{0.8}\,\mathrm{d}\,\dfrac{\{1-(d_1/d)\}}{\{1+(d_1/d)^{0.2}\}}$

where d = diameter of the drill (in mm), d_1 = diameter of the hole to be enlarged,

k = constant depending on the number of flutes, which can vary from 0.87 for single flute to 1.59 for 20 flutes in a reamer. For an eight-flute reamer, which is commonly used, k will be 1.32.

A.3.3 Taps

These are basically helical screws having flutes cut along its axis, allowing cut edges of the threads to act as cutting tools. The taps have chamfer angle at their bottom, whose length will be nominally 11/4, 4 and 8 thread pitches. These are called "bottoming", "plug" and "taper" taps, the first two being for roughing operation and the last one for finishing operation. The flutes could either be straight or spiraled. Although the taps are hand-operated, many of the taps could be operated through machine. Therefore, these could be used in drilling machines once the drilling operations are completed by using different sized bushes. However, the speed of rotation will be considerably reduced. Similar to single-point cutting tools, the cutting edges of taps could be ground with the required rake angles. Geometrical details of the taps are given in Figs A.8 and A.9. While Fig. A.8 gives the overall nomenclature, Fig. A.9 provides the details of the different kinds of cutting edges and their profiles. Generally, taps are made of HSS. In certain cases, they could be cemented carbides made by the powder metallurgy route required for hard abrasive materials.

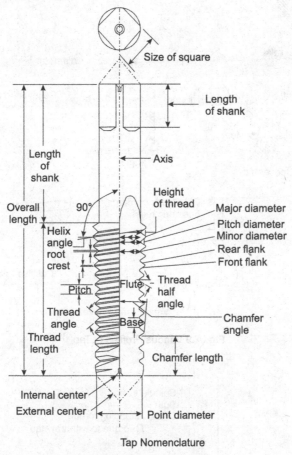

Tap Nomenclature

Fig. A.8 Nomenclature of Tap

Forces that act during the tapping operation consist of mainly the torque. The thrust, as in the case of drilling, is minimal. Therefore, the clamps for this operation mainly consist of restraining forces during operation and during sudden failure of tap.

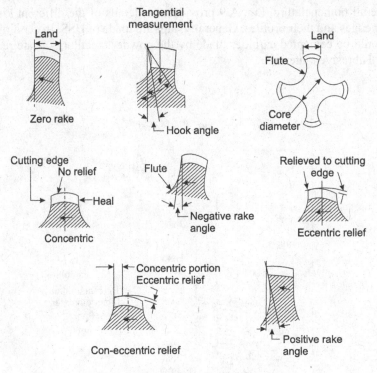

Various Tap Profiles

Fig. A.9 Various Profiles of Tap

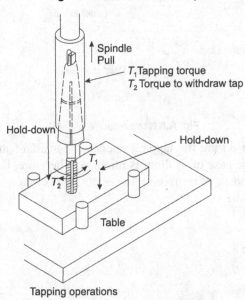

Tapping operations

Fig. A.10 Clamping Forces Needed to Restrain the Component

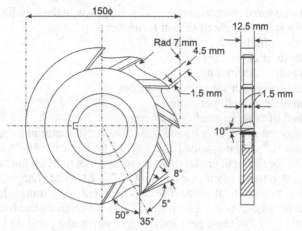

Typical 150 mm diameter, 35°rake angle milling cutter

Fig. A.11 Typical View of a Peripheral Milling Cutter

A.3.4 Milling Cutters

Milling processes can be broadly classified into two types, *viz.* (*a*) Face milling and (*b*) Peripheral milling. In the face milling operation, the metal gets removed at the surface parallel to the milling cutter. One example is the milling of counter-sunk hole to accommodate for screws in the assembly. In the peripheral milling operation, the metal gets removed at the surface parallel to the circumferential area of the cutter (Fig. A.11). Examples are (*i*) keyway milling, and (*ii*) making slots in different shaped components. Peripheral milling can be further classified into up-cut milling and down-cut milling. In up-cut milling, the feed or the movement of the component is in the opposite direction to the peripheral velocity of the cutter. Here, the thickness of the metal removed in each cutter rotation increases as the component advances. In the case of down-cut milling, the feed and the peripheral velocity of the cutter are one and the same. The thickness of metal removed is the highest when the cutter touches the component and progressively reduces as the feed increases.

The profile of a milling cutter tooth resembles a single-point cutting tool as the tooth form has rake angles (both front and side rake angles), top and side lands, primary and secondary clearances as well as side clearances. In the case of face milling cutters, the cutter tooth takes the shape of a spiral along the width of the cutter, thus making it possible to be used for metal removal along its width as well. In the modern production shops, the milling cutters have throwaway tips also, so that cemented carbide tips can be used for high production. Negative rake angles are widely used these days for increased productivity.

As regards the force calculations, initially the power at the spindle is calculated. The same is given by the rule of thumb formula:

Power at the spindle in $h_p = d\,W\,N\,n\,f\,C$

where d = depth of cut in mm

         W = width of cut in mm

         N = revolution per minute of the cutter

         n = number of teeth per cutter

         f = feed of table in mm/tooth of cutter

         c = constant depending on machine condition, material to be cut, etc.

The constant "C" can be assumed as 4.3×10^{-5} for brass, 6×10^{-5} for cast iron, 8×10^{-5} to 12×10^{-5} for different grades of steel proportional to its hardness. If the power is to be expressed in terms of kW, the constant "C" needs to be changed accordingly.

Thus, for a hypothetical data, like depth of cut of 2 mm, width of cut of 12 mm, speed of arbour being 200 rpm, 12 nos. of teeth as the total number of teeth in the cutter, feed of table per tooth as 0.30 mm and a constant C of 12×10^{-5} (assuming hard steel), the power at the spindle can be calculated as 2.1 h_p.

This could also be derived from the power of the motor provided in the machine. If the motor power in a machine is 5 h_p, and the efficiency of the drive system is 52%, taking into account the gears, friction in the bearings, etc., the power available at the arbour will be 2.6 h_p.

Another method for the determination of the power at the arbour is as follows:

     Power required at the cutter = Metal removal rate/ Constant K.

Where metal removal rate can be expressed in terms of cm^3/min and the constant K depends on various factors like material hardness, feed per tooth, and the thickness of the metal being removed.

As per the standard codes, the force in cutting direction can also determined from the spindle h_p, and the formula is applicable for the tools which travel in a linear fashion:

Cutting Force F in Newton = $(45000 \times h_p)$/ Cutting Speed in m per min

As the determination of cutting forces are quite complex due to the number of variables such as cutting speed, feed rate, depth of cut, number of teeth, material hardness, cutting conditions, etc., careful examination of the selection and application of clamps are necessary. In addition to normal forces, regenerative process of chatter of the cutter may cause undue stresses at the clamps. Deployment of hold-down bolts, tenons and rugged fixture body to dampen the dynamic forces are essential requirements of fixture design.

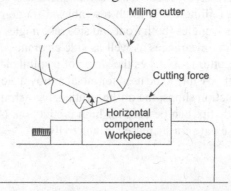

Fig. A.12 Cutting Forces in Milling Operation

A.3.5 Broaching Tools

Broaching is a process of metal removal through a linear movement of cutting tool which has independent cutters placed in sequence. Each successive cutter tooth will be larger in size than the preceding one by an amount equal to the volume of metal removal possible by each tooth. Thus, each broaching tool caters to only a specific application of metal cutting, which could be either internal or external surface of a component. Most widely used application is the machining of internal splines and internal keyways. Being job-specific, the rate of production in broaching operation is quite high. However, the cost of this special-purpose machine should warrant for the high production rate. Broaching tool has profiles like a milling cutter, with rake angles and primary and secondary relief angles. Generally, the tool is made up of high-speed steel. Unlike a milling fixture, the broaching fixture does not need elaborate clamping system as the linear cutting force acts like a clamping force and holds the component to the collar of the machine. The component is held on to the machine through a support plate shaped to suit the component and the machine opening. Figures A.13 and A.14 give details of the profiles of a broaching tool and the method of holding to the machine, respectively.

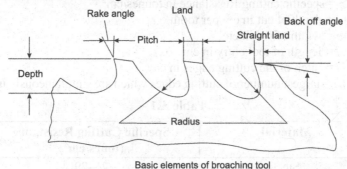

Basic elements of broaching tool

Fig. A.13 Geometry of Broaching Tool

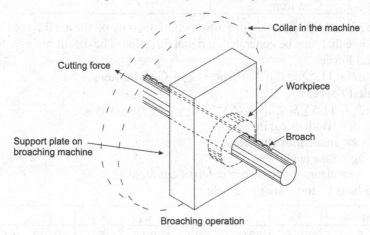

Broaching operation

Fig. A.14 Method of Holding Components in Broaching

Forces in Broaching Operation

In this section, the numerical methods which are adopted for the calculation of broaching force are discussed. Figure A.15 shows the principle of broaching operation. The cutting edges are arithmetically stepped and hence the cutting resistance exerted on all cutting edges are approximately equal. Thus, the resultant broaching force acting on the broach varies with the approaching length. The principle is explained in Fig. A.16. It can be seen that L represents the total length of the machined surface, while l represents the length over which the largest number of cutting edges are engaged on the component, and l represents the total length during which the chip section remains constant. t signifies the pitch of the broaching tool and P denotes the total axial force at a given length of tool contact with the workpiece.

There are two formulae to evaluate the broaching force. The first one being as follows:

Maximum broaching force (in tonnes), $P = k\,K_s abl/t$

where k = functional constant which varies between 1.1 and 1.3

Ks = specific cutting Resistance in tonnes/cm^2

a = depth of cut in cm per tooth

b = width of cut in cm

l = length of tool entry in cm

t = pitch of the cutting edges in cm

$1/t$ = largest number of cutting edges which are simultaneously in action.

Table A.1

Material	Specific Cutting Resistance (tones/cm^2)
Steel	21–29
Al–Si alloy	10.4
Cast iron	19

The second formula provides information not only on the axial force but also on the force that may be generated in radial direction. The details are as follows:

Axial Force:

$$P = 11.5\sum b(C_1 a^{0.85} + C_2 k - C_3\gamma - C_4\alpha)\text{ newtons.}$$

Radial Force:

$$P_a = 11.5\sum b(C_5 a^{1.2} + C_6 a^{1.2} + C_6\gamma - C_7\alpha)\text{ newtons.}$$

where b = width of cut in mm

k = number of chip breakers

g = rake angle

a = clearance angle α = depth of cut/tooth

The constants C_1 to C_7 are given below:

Material	C_1	C_2	C_3	C_4	C_5	C_6	C_7
0.2% C Steel	115	0.06	0.2	0.12	55	0.018	0.045
0.45% C Steel	220	0.108	0.32	0.14	215	0.081	0.117

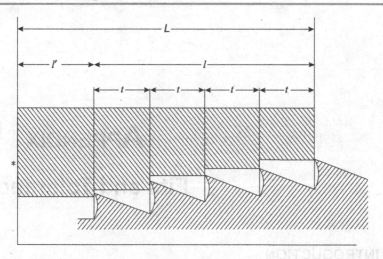

Fig. A.15 Cutting Action of Broach Tool with Arithmetically Progressing Teeth

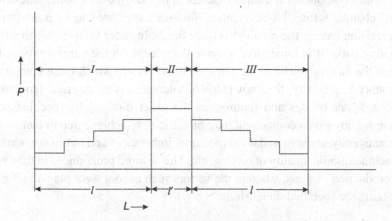

Fig. A.16 Axial Force diagram in Broaching showing the
Progression of Load as the Cutter Moves

APPENDIX $\boxed{\text{B}}$

Fits and Tolerances

B.1 INTRODUCTION

Many of the components in the manufacture of jigs and fixtures, such as the drilling bushes, clamps, setting blocks, tenons, fasteners, etc. need to be manufactured with precision having the required surface finish. In order to assemble them with the mating parts, it is imperative to have knowledge on the various types of fits between the mating pairs and the required tolerances to which these components are produced. Secondly, the knowledge of tolerances is an essential requirement by the designer of jigs and fixtures, as the exact dimensions specified in the drawing for any given component may be difficult to achieve due to factors such as inhomogenity in the material composition, imbalances and out-of-roundness in the machine spindle, quality of tooling, etc. This is more pronounced in the case of mass production of parts, wherein the factors such as tool wear play a major role in retaining the specified dimensions.

B.2 UNILATERAL AND BILATERAL TOLERANCES

There are basically two types of mentioning the tolerances (upper limit and lower limit), namely, unilateral and bilateral. In the case of unilateral tolerances, the tolerance is specified in either positive or negative with respect to base dimension of the component. For example, a shaft diameter can be dimensioned as $\phi\ 20^{+0.2}_{-0.0}$ or a hole can be dimensioned as $\phi\ 20^{+\ 0.0}_{-\ 0.02}$. In both these cases, increase either in positive or in negative direction only is permitted while the respective component is machined. In other words, the shaft can vary in its diameter between 20 and 20.2. Similarly, the hole can have its diameter between 25 and 24.98.

Bilateral tolerance can be done on both positive and negative directions, such as $30^{+0.4}_{-0.2}$, specifying that a dimension of the component can vary between 29.8 and 30.4 after machining.

Design of Jigs, Fixtures and Press Tools, First Edition. K. Venkataraman.
© K. Venkataraman 2015. Published by Athena Academic Ltd and John Wiley & Sons Ltd.

B.3   SHAFT AND HOLE BASIS OF SPECIFYING TOLERANCES

In case of specifying the tolerances of mating parts, either Shaft Basis or Hole Basis systems are followed. In the case of Shaft Basis system, the dimension of the shaft is retained as constant and the hole dimension for various types of fits are varied. Basic shaft basis of indicating the tolerances is shown in Fig. B.1 and the hole basis of indicating tolerances is explained in Fig. B.2. Both these systems have three types of fits: (*a*) Positive allowances, otherwise known as clearances; (*b*) Negative allowances, otherwise known as interferences; and (*c*) Zero allowance, which is not generally adopted in engineering applications, and is only a hypothetical representation.

In the system of "Basic Hole", the dimension of the hole is maintained constant, while the shaft diameter is varied to meet the various fits required. Since machining or varying the dimensions of shaft is easier than changing the diameter of the hole, the later one, *viz.* "Basic Hole" system is adopted universally now.

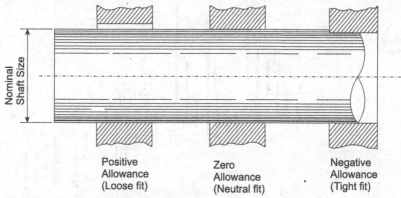

Positive   Zero   Negative
Allowance   Allowance   Allowance
(Loose fit)   (Neutral fit)   (Tight fit)

Fig. B.1 Basic Shaft System in Specifying the Tolerances

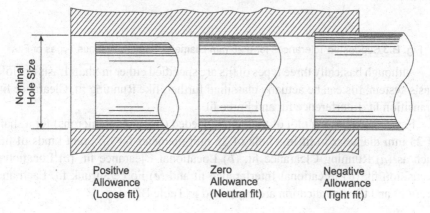

Positive   Zero   Negative
Allowance   Allowance   Allowance
(Loose fit)   (Neutral fit)   (Tight fit)

Fig. B.2 Basic Hole System in Specifying the Tolerances

As the type of machining has a bearing on the tolerances that could be achieved, International Standards as well as Indian Standards Institutions have specified the tolerances that need to be maintained for various diameters of holes. Table B.1 indicates such tolerances to be maintained for different notations like H_7, H_8, H_9 and H_{11}. A notation of H_7 indicates that the machined hole is produced after reaming, whereas H_{11} indicates a hole produced after drilling and the intermediate notations indicate the type of surface finish the selected machining process will produce and the tolerances to be applied for.

Table B.2 indicates the tolerances to be maintained for shafts of different diameters. A notation of p_6 or s_6 may indicate a required tolerance for a force fit or an interference fit, whereas a notation of f_7 may indicate a clearance or a running fit. Specification of a correct notation is needed to be done judiciously for the required application.

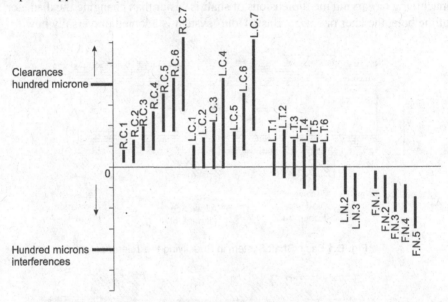

Fig. B.3 Suggested Tolerances for a 25 mm Diameter Shaft for Various Types of Fits

Although basically three types of fits are specified either in Shaft Basis or Hole Basis system, fits can be actually classified further, like Running fit, Clearance fit, Transition fit, Interference fit, and Force fit.

Figure B.3 shows a plot of tolerances needed (in hundred microns) for a shaft of 25 mm diameter to fit in a hole of 25 mm diameter for different kinds of fits such as (*a*) Running Clearance fit, (*b*) Locational Clearance fit, (*c*) Locational Transition fit, (*d*) Locational Interference fit and (*e*) Force/Shrunk fit. Each and every fit and their application are explained in Table B.3.

Table B.1 Tolerance Limits for Selected Holes (Hole Basis)

Nominal Sizes		H_7		H_8		H_9		H_{11}	
Over (mm)	Upto and including (mm)	Upper Limit +	Lower Limit	Upper Limit +	Lower Limit	Upper Limit +	Lower Limit	Upper Limit +	Lower Limit
6	10	15	0	22	0	36	0	90	0
10	18	18	0	27	0	43	0	110	0
18	30	21	0	33	0	52	0	130	0
30	50	25	0	39	0	62	0	160	0
50	80	30	0	46	0	74	0	190	0
80	120	35	0	54	0	87	0	220	0
120	180	40	0	63	0	100	0	250	0
180	250	46	0	72	0	115	0	290	0

Unit = 0.001 mm

Table B.2 Tolerance Limits for Selected Shafts (Shaft Basis)

Nominal Sizes		c_{11}		d_{10}		e_9		f_7		g_6	
Over to (mm) −		Upper Limit −	Lower Limit −	Upper Limit −	Lower Limit −	Upper Limit −	Lower Limit −	Upper Limit −	Lower Limit −	Upper Limit −	Lower Limit −
6	10	80	170	40	98	25	61	13	28	5	14
10	18	95	205	50	120	32	75	16	34	6	17
18	30	110	240	65	149	40	92	20	41	7	20
30	40	120	280								
40	50	130	290	80	180	50	112	25	50	9	25
50	65	140	330								
65	80	150	340	100	220	60	134	30	60	10	29
80	100	170	390								
100	120	180	400	120	260	72	159	36	71	12	34
120	140	200	450								
140	160	210	460	145	305	85	185	43	83	14	39
160	180	230	480								
180	200	240	530								
200	225	260	550								
225	250	280	570	170	355	100	215	50	96	15	44

Unit = 0.001 mm

Nominal Sizes Over to (mm)		h_6		k_6		n_6		p_6		s_6	
		Upper Limit −	Lower Limit −	Upper Limit +	Lower Limit +	Upper Limit +	Lower Limit +	Upper Limit +	Lower Limit +	Upper Limit +	Lower Limit +
6	10	0	9	10	1	19	10	24	15	32	23
10	18	0	11	12	1	23	12	29	18	39	28
18	30	0	13	15	2	28	15	35	22	48	35
30	40	0	16	18	2	33	17	42	26	59	43
40	50										
50	65	0	19	21	2	39	20	51	32	72	53
65	80									78	59
80	100	0	22	25	3	45	23	59	37	93	71
100	120									101	79
120	140									117	92
										125	
140	160	0	25	28	3	52	27	68	43	133	100
160	180									151	108
180	200									159	122
										169	130
200	225	0	29	33	4	60	31	79	50		140
225	250										

Unit = 0.001 mm

Table B.3 Various Types of Fits and Their Applications

S. no.	Types of Fits	Applications
1	Running fit (RC)	RC1: Close sliding fits intended for accurate location of parts which will not have any play. Example: Anti-friction bearings in household appliances. RC2: Sliding fits intended for accurate location, but with greater clearance than in RC1. Example: Clearances for drill bit movement inside a guiding bush. RC3: Close running fits intended for slow speeds at light loads. Example: Bearings in mechanical press work. RC4: Close running fits intended for slow speeds at medium loads. Example: Machine tool bearings. RC5 and RC6: Medium running fits intended for higher speeds at heavier loads. Example: Motor bearing transmitting heavy torques to gear drives for reduction of speeds. RC7: Intended for free running fits when accuracy is not essential and larger temperature variation is anticipated. Example: Babbitt coupling; fabric bearings in rolling mills.
2	Locational Clearance fits	These are usually applicable for stationery parts, which are assembled and dis-assembled. Example: Indexing pin and groove/slot located in the index plate.
3	Locational Transition fits	These have both clearance and transition fits, and are used when location is important but certain clearance is permitted. Example: Sliding 'V' clamps moving over a dovetail joint.

...Contd.

4	Locational Interference fits	These are applicable when accuracy of location is the prime important factor. Example: Cylindrical locators fitted to the jig body frame.
5	Force fit/ Shrunk fit	FN1: Requires light assembly pressure. Example: Assembly of Light bearing of upto 25 mm diameter. FN2: Requires medium assembly pressure used for steel parts. Example: Assembly of bearings in large diameter shafts; bearings for sizes exceeding 100 mm diameter; and drill bush in Jigs. FN3: Requires heavy driving force. Example: Driving a wedge for moving heavy machinery. FN4 and FN5: Require heavy assembly pressure through presswork operated hydraulically. Example: Railway wheels and axles pressed hydraulically.

Figures B.4, B.5 and B.6 show the clearances and interferences (in hundred microns) needed for various nominal diameters of shafts.

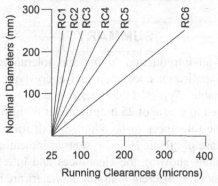

Fig. B.4 Running Fit for Various Nominal Diameters

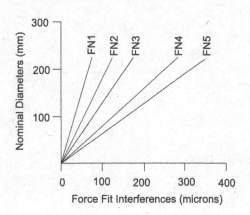

Fig. B.5 Force/shrunk Fit for Various Nominal Diameters
(interferences in hundred microns)

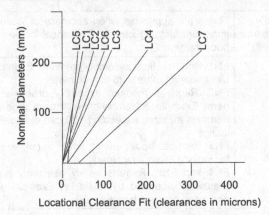

Fig. B.6 Locational Clearance Fit for Various Nominal Diameters
(clearances in hundred microns)

SUMMARY

In this appendix, an introduction to fits and tolerances is given with explanations and applications for each and every type of fit. This has been shown as a table. Typical tolerances needed for a shaft of 25 mm diameter when fitted to a hole of 25 mm diameter is shown as a bar chart. Tables showing the tolerances to be adopted both for a hole as well as for a shaft in the case of "Basic Hole" system of indicating the tolerances are presented. Graphs showing the clearances and interferences needed (in hundred microns) for different nominal diameters are illustrated.

❑❑❑

APPENDIX $\boxed{C}$

Jigs and Fixtures: Suggested Questions and Answers

Q. 1. **What is the difference between a drill jig and a fixture?**

Ans. A drill jig locates and clamps the workpiece as well as guides the drilling tool; whereas a fixture merely locates and clamps the component. However, a milling fixture identifies the cutter with respect to the component clamped for machining by means of a setting block.

Q. 2. **Distinguish between a drill jig and a drill fixture.**

Ans. A drill jig, in addition to locating and clamping the component, guides the tool as well. In the case of drill fixture, guidance of the tool does not take place, as there is no provision of guide bushes. Such fixtures are generally used for rolled sections requiring fabrication. Accuracy of the coordinates of the hole to be drilled is not of prime importance.

Q. 3. **What forces does a drill bit exert on a workpiece?**

Ans. The forces that a drill bit exert on a workpiece are:

- Thrust in vertical direction taken over by the table. Theoretically, no clamping forces are needed to resist such thrust.

- Torque due to flutes in the drill bit and the same is resisted by constraints provided in the opposite direction to rotation. In case of a breakage of the drill bit while drilling, an opposite vertical force to the thrust will try to lift the component. Actually, clamping helps in preventing such lifts.

Q. 4. **Is it correct to locate drill bushes in the leaf of a leaf-type drill jig?**

Ans. No, as this results in errors in verticality of the drill bushes.

Q. 5. **State the reasons for avoiding the clamping of a component with a hinged plate of a leaf-type jig?**

Ans. Many times the hinges become vulnerable against the cutting and clamping forces and may create maintenance problems.

Q. 6. **Why small projections that project outside the drill jig are avoided?**

Ans. Small projections that project outside the drill jigs are avoided:

- So that such projections do not foul with the movement of components, while loading and unloading.
- To avoid physical injury to the operator.

Q. 7. **"Complex clamping devices are avoided". State True or False. Give reasons for your answer.**

Ans. True. Complex clamping devices are avoided because such systems are expensive and require skill in maintenance.

Q. 8. **"The locating points in a drill jig should be visible." State True or False. State the reasons for your answer.**

Ans. Yes. This will enable quicker and accurate positioning of the component before clamping. However, the same may not be feasible in all the cases. Adoption of the technique of "foolproofing" 'which makes it possible to load the component to the jig in only one unique way, may enable the operator to locate without visibility of the locating pins.

Q. 9. **What is the distance between the bottom of a drill bush and the workpiece?**

Ans. The distance between the bottom of a drill bush and the workpiece is 1.5 times the diameter of the hole drilled, and this is provided to enable chip removal.

Q. 10. **What are the methods employed for removal of chips from the jig?**

Ans. The methods employed for removal of chips from the jig are:

- Manual brushing of the jig area under the component, once the component is unclamped and removed
- By properly directing the compressed air so that the chips do not get entangled in bearings and joints of the machine tool
- By the flow of cutting solvents.

Q. 11. **What is the general rule to decide the length of a drill bush?**

Ans. The length of a drill bush should be twice the diameter.

Q. 12. **Generally, it is a practice to have four legs for the drill jig instead of three. Give your explanation.**

Ans. It is a practice to have four legs, instead of three, for the drill jig to provide for better stability of the jig, especially when subjected to large thrust of drill bit.

Q. 13. **Why stress relieving is done for the jig bodies, which are fabricated through welding process?**

Ans. Stress relieving is done for the jig bodies fabricated through welding to relieve the thermal stresses retained while welding which may cause distortions.

Q. 14. **What is the hardness of a drill bush?**

Ans. Hardness of 60–64 Rc can be used.

Q. 15. **What are the main advantages for choosing cast-constructed jig body?**

Ans. The main advantages are: (*a*) complicated shapes can be cast and (*b*) such bodies absorb heavy vibration and chatter while drilling operation.

Q. 16. **What are principal and secondary locators? Explain with example.**

Ans. Principal locators are generally cylindrical in shape and can locate the main hole in a component having two holes to be located. An example is the locating of a connecting rod having big-end and small-end holes. The secondary hole is located by means of a diamond pin locator which is not cylindrical but has six sides. Two of its sides are curved, whose profile represents the lower limit of the diameter of the second hole. The other four sides are formed by relieving the cylinder and are of straight sides. Such locators are used when the center distance between the two holes (as in the case of the connecting rod) can vary as per the specified tolerances.

Q. 17. **What are the disadvantages of nesting type locators?**

Ans. Nesting type locators are used for specific components having definite profiles and therefore cannot be used even for other types of components having similar profiles. Secondly, if the profiles are complicated, the cost of machining such profiles will be expensive.

Q. 18. **If two or more holes to be drilled are very close, what is the method used in providing bushing?**

Ans. A single bush having both the holes adjacently will be provided.

Q. 19. **What is the clearance provided between the drill bush and the drill bit?**

Ans. The clearance provided between the drill bush and the drill bit ranges between 0.001 and 0.02 mm.

Q. 20. **What are the general classifications of drill bushes?**

Ans. The general classifications of drill bushes are given as follows:

- Plain Liner bush
- Collared bush
- Replaceable Collar bush
- Screw-type bush
- Locating and Clamping bush

Q. 21. **Why should the drill feet be ground after the assembly with the jig frame, rather than before the assembly?**

Ans. The drill feet should be ground after the assembly with the jig frame as this will enable squareness of the surfaces ground with respect to the horizontal plane. In addition, this method will ensure accuracy in angular position of the drill bushes as the height of all the jig feet could be ensured to be uniform.

Q. 22. **What is the type of fit between a jig body and a liner bush? How is a liner bush of outside diameter of 20 mm dimensioned in a drawing to take care of the tolerances? How does the hole in the jig frame into which the liner bush is fitted, is dimensioned?**

Ans. The type of fit between a jig body and a liner bush is Interference fit. The liner bush is dimensioned as $\phi\,20p_6$. The hole is dimensioned as $\phi\,20H_7$.

Q. 23. **What is the type of jig used for drilling a number of holes in more than one orthogonal plane of a component?**

Ans. Box type jigs are used for drilling a number of holes in more than one orthogonal plane of a component.

Q. 24. **How does an indexing jig work?**

Ans. Indexing jig has basically two main components:

(*a*) Circular index plate with either slots or grooves placed equally (or) as per the requirement in the component. Index plate is allowed to rotate about an axis and the rotation is effected by a knob fitted to the index plate.

(*b*) Indexing mechanism consists of spring-actuated pin or ball which is designed to engage onto the slot/groove of the index plate effecting the indexing operation in discrete steps. Each time when the indexing operation is complete, the drilling is effected. The cycle is repeated until the entire drilling operations are over.

Q. 25. **What are the cost considerations in the use of a fixture?**

Ans. The cost of manufacture of a fixture and its maintenance cost to produce components of required accuracy, should be lesser than the cost saved due to productivity gain by using the fixture.

Q. 26. **Why are keys/tenons mounted at the base of the milling fixture?**

Ans. These are provided so as to locate the fixture with respect to the machine table.

Q. 27. **What are the various types of production milling?**

Ans. The various types of production milling are:

- Plain milling
- Straddle milling
- String milling
- Gang milling
- Indexing milling

Q. 28. What are the major advantages of vacuum and magnetic fixtures?

Ans.

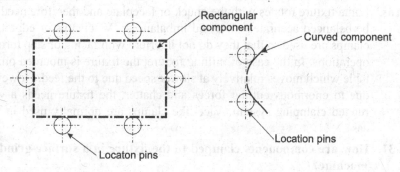

Fig. C.28.1. Vacuum Chucking System Used for Flat Thin Components, both Ferrous and Non-ferrous

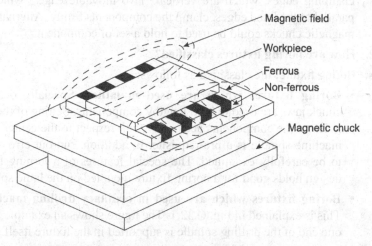

Fig. C.28.2. Magnetic Chucking of Ferrous Components

Vacuum chucking helps in holding ferrous as well as non-ferrous components, particularly strips and plates of large surface area. The same is explained in Fig. C.28.1. Vaccum chucking is quick in operation and can exercise uniform clamping force throughout the area. Magnetic chucks (work on electromagnetic force) are used mainly in surface grinding of ferrous components. It is quite rapid in holding and release of components. The disadvantage of such clamping is that the same cannot be used for non-ferrous components. Figure C.28.2 explains such chucks.

Q. 29. What are the classifications of broaching fixtures?

Ans. Broaching fixtures are classified into:

  • Internal

  • External

Q. 30. What are the basic differences between a lathe fixture and a milling fixture?

Ans. Lathe fixture rotates with the chuck or faceplate and therefore needs to be balanced accurately to avoid imbalances. 'V' clamps or edge-type clamps are used so that they do not interfere with facing or step turning operations. In the case of milling fixture, the fixture is mounted on the table which moves relatively at slower speed due to the feed. However, due to enormous cutting forces and chatter, the fixture needs a very rugged clamping system. Vice-like clamps are normally used in this case.

Q. 31. How are components clamped to the fixture in a surface-grinding machine?

Ans. Batch of components are placed on a horizontal plane having two fixed clamping edges, which are vertical. Two movable edges, which are parallel to the fixed edges, clamp the components firmly. Alternatively, magnetic chucks could be used to hold a set of components.

Q. 32. How are boring fixtures classified?

Ans. Boring fixtures are classified as follows:

- **Boring fixtures which are used in lathes:** Specially designed chuck jaws are used to hold lengthy components. Setting of the pilot holes of the components accurately with respect to the center of the machine spindle is a pre-requisite. In addition, run-out errors need to be carefully examined. The special features of a turning fixture design holds good for a boring fixture mounted on the lathe spindle.

- **Boring fixtures which are used in ordinary drilling machines:** This is explained in Fig. C.32.1. The figure shows an example where one end of the drilling spindle is supported in the fixture itself.

- **Boring fixtures which are used in horizontal boring machines:** Figures C.32.2 and C.32.3 explain such varieties. One type has a support of the boring bar on both the ends, whereas in the other one, the boring bar is supported on a bushing located on the component itself.

- **Boring fixtures which are used in heavy vertical boring machines like railway wheel boring machines:** These require locating the component accurately with respect to the chuck or working holding devices, which rotate in horizontal plane. Therefore, balancing of the rotating member is not a major issue in these fixtures.

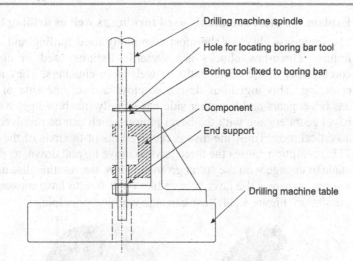

Fig. C.32.1 Boring Fixture Mounted on a Drilling Machine Table

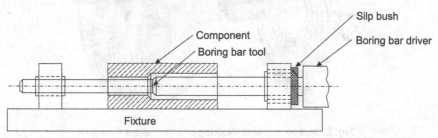

Fig. C.33.2 Boring Fixture Mounted on a Horizontal Boring Machine with Dual Supports

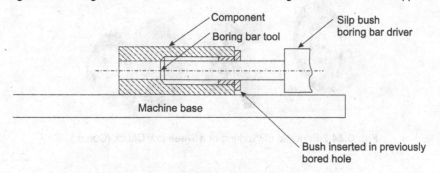

Fig. C.33.2 Boring Fixture in a Horizontal Boring Machine with
Support in the Component itself

Q. 33. What is the method of locating a fixture onto the faceplate of a lathe?

Ans. The methods of locating a fixture onto the faceplate of a lathe are:

- By dowel pins and screws
- By having suitable brackets to locate and support the fixture.

Q. 34. **Explain the most commonly used turning as well as drilling fixture.**

Ans. The three-jaw chuck is the most commonly used turning and drilling fixture. Three-jaw chucks are versatile fixtures used in clamping components in the center lathe as well as in chuckers. They are self centering. This ingenious design employs a disc, one side of which has bevel gears and the other side has spirally machined grooves. The bevel gears engage with the bevel pinion which can be revolved about a vertical axis. Thus, the disc is revolved about the axis of the chuck. This revolution causes the three jaws to move up and down, as they are made to engage with the spiral grooves. Thus, the rotating disc acts like a screw, whereas the jaws behave like a nut free to have movement in translation. Figure C.34.1 explains the working principle.

Fig. C.34.1 Principle of Working of a Three-jaw Chuck

Fig. C.34.2 Principle of Working of a Three-jaw Chuck (Contd.).

❑❑❑

PART-II

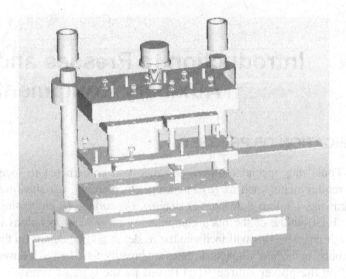

PRESS TOOLS

Introduction to Presses and Auxiliary Equipment

1.1 CLASSIFICATION OF PRESSES

Although 'Press Tools' are special toolings designed and constructed to meet specific forming requirements such as (*a*) blanking, (*b*) piercing, (*c*) shearing, (*d*) bending, (*e*) drawing, (*f*) deep drawing, (*g*) shaving, (*h*) curling, (*i*) embossing, (*j*) coining, etc., a brief outline of the main equipment used in such operations is explained here in this chapter. This will facilitate the reader to go into details on the subject of "Press Tools" in the later chapters. Presses are broadly classified as follows:

   (*i*) Based on the power source   (*ii*) Based on the type of frame
   (*iii*) Based on the method of actuation of slides
   (*iv*) Based on the number of slides in action

1.2 CLASSIFICATION BASED ON POWER SOURCE

Based on the power source, presses can be classified as:

   (*a*) Mechanical presses   (*b*) Hydraulic presses.

   Table 1.1 makes a comparative analysis of the above two types.

Table 1.1 Comparative Analysis of Mechanical and Hydraulic Presses

S. No.	Characteristics	Comparison of Mechanical and Hydraulic Presses	
		Mechanical	**Hydraulic**
1.	Force	Varies accordingly to slide position	Constant
2.	Capacity	50 MN (6000 tonnes)	440 MN (52800 tonnes)
3.	Stroke length	Limited	Capable of 2.5 m stroke
4.	Slide speed	Higher than hydraulic presses; maximum at mid stroke	Slower pressing speeds; uniform speed throughout the stroke
5.	Control	Full stroke is to be effected before reversal	Slide can be reversed at any position
6.	Application	Used for large production rates such as in blanking and piercing operations	Used in deep drawing, drawing of irregular shapes, straightening, and operations requiring variable or partial strokes

Design of Jigs, Fixtures and Press Tools, First Edition. K. Venkataraman.
© K. Venkataraman 2015. Published by Athena Academic Ltd and John Wiley & Sons Ltd.

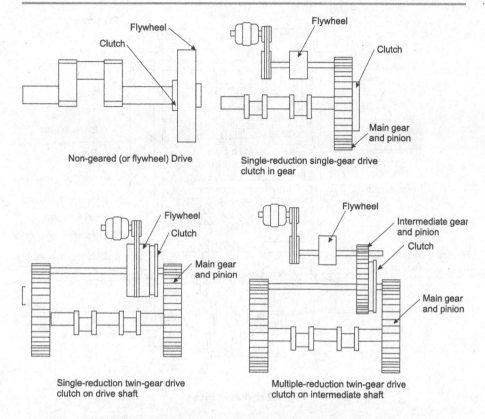

Fig. 1.1 Different Types of Drives in Mechanically Driven Presses

Figure 1.1 shows the types of drives used in mechanical presses. In most mechanical presses, a flywheel is the major source of energy that is applied to the slides by cranks, gears, eccentric or linkages. The flywheel runs continuously and is engaged by the clutch only when a press stroke is needed.

Figure 1.2 shows a typical hydraulic press. Basic operation of a hydraulic press consists of applying hydraulic pressure at one end of a piston sliding inside a cylinder. Thus, the hydraulic press can be used to apply an enormous force of up to 50,000 tonnes. The press can also deploy longer strokes as well as variable strokes enabling deep drawing operations. Hydraulic presses generally have the following items to make up for the complete system:

    (*i*) High-pressure hydraulic pump   (*ii*) Accumulator

  (*iii*) Oil storage tank   (*iv*) Solenoid actuated values

    (*v*) Cylinders and piston assemblies (separate cylinder and piston assemblies are used for the main punching operation and for the die cushion assembly)

  (*vi*) Pressure switches

 (*vii*) Pressure gauges

(*viii*) Filter

  (*ix*) Interconnecting pipeline

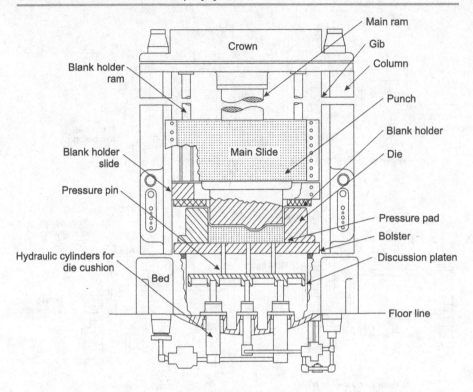

Fig. 1.2 Hydraulic Press

1.3 CLASSIFICATION BASED ON TYPE OF PRESS FRAME

Based on the type of press frame, presses can be classified as:

    (*a*) Gap-frame presses   (*b*) Straight-side presses.

Gap-frame presses are sometimes called C-frame presses because the frame resembles the letter 'C' when viewed from the side. Gap-frame presses can be further classified into:

    (*i*) Open back inclinable   (*ii*) Bench press

    (*iii*) Adjustable bed stationary (*iv*) Open back stationary.

The advantages of gap-frame presses are that the strip can be fed from an un-coiler located at one side of the gap and the coiler can be located on the other side of the gap, enabling the movement of stock with ease. The finished component can be pushed out towards the rear of the press.

The first type of gap-frame press, *viz.* "open back inclinable" enables the press frame, including the bed, to be inclined with respect to its base frame, causing the finished component to slide out by gravity. However, the main disadvantage of such presses is that the overload causes misalignment of dies and punches, resulting in die wear at an early stage itself. Figure 1.3 shows a typical gap-frame press.

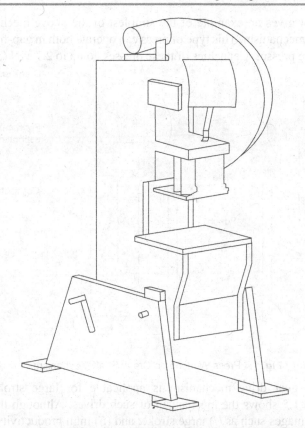

Fig. 1.3 Gap-frame Press (open back inclinable)

A straight-side press has a frame made up of base/bed, two columns, and a top crown. The straight-side presses are basically mechanical presses that work on the principle of eccentric shaft or eccentric gear drive. In addition to the drive system, these presses have pneumatically actuated counter-balance cylinders to assist in pushing the slide and the upper die to the top of the stroke. They also facilitate smooth press operation and easy slide adjustment. These presses are also provided with die cushion cylinder, which assist in holding the blank before the operation and ejection of the finished product from the die. Straight-side presses have capacities ranging from 180 kN to 36 MN. The main advantage of this press over the gap-frame press is that the same deflects less under off-center loads and hence the wear of the die is minimal.

1.4   CLASSIFICATION BASED ON METHOD OF ACTUATION OF SLIDE

The most widely used mechanisms in operation of these presses are:
    (*a*) Crank shafts   (*b*) Eccentric shafts
    (*c*) Knuckle lever drives   (*d*) Rocker arm drives
    (*e*) Toggle mechanisms.

Figure 1.4 gives an example of the simplest of the above mechanisms, *viz.* "crank shaft" mechanism. This type of press can operate both in gap-frame as well as straight-side presses. Capacities of these presses go up to 2.7 MN (300 tonnes).

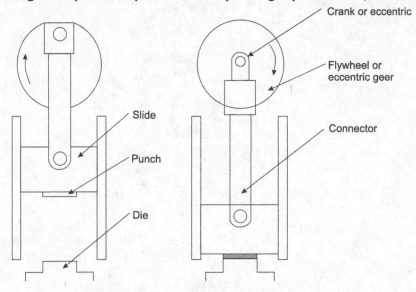

Fig. 1.4 Press Working on Crank Shaft Mechanism

Eccentric gear drive mechanism is applicable for large strokes of upto 1.3 m. Figure 1.5 shows the mechanism of such drives. Although these presses have the advantages such as (*i*) large stroke, and (*ii*) high productivity; they have the disadvantages such as (*i*) over-hung flywheel, (*ii*) more cost for single-gear eccentric press (due to the gear systems involved) and (*iii*) higher probability of striking at the bottom due to greater friction at the joints.

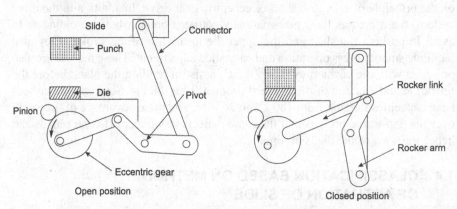

Fig. 1.5 Eccentric Gear Type Presses

1.5　CLASSIFICATION BASED ON THE NUMBER OF SLIDES IN ACTION

Based on number of slides, presses can be classified as (*a*) one-slide, (*b*) two-slide, and (*c*) three-slide presses, which are referred to as single-, double- and triple-action presses.

Single-action press has one reciprocating slide acting against a fixed bed. Generally, these type of presses are used in metal stamping operations such as blanking, embossing, coining and drawing.

Double-action press has two reciprocating slides acting in the same direction. These are mainly used in deep drawing operations, wherein the first slide having a shorter stroke called the blank holder slide moves first. Once the blank is firmly gripped by the blank holder (which is hollow inside, and has an annular surface all round), the punch mounted on to the second slide and moves inside the hollow blank holder, carrying out the drawing operation by depressing the cushion provided in the die.

Triple-action press has two slides working in the same direction as in the case of double-action press. In addition, the third slide works in the opposite direction, opposite to the blank holder and inner slides. This operation facilitates reverse drawing, forming or beading operation. Figure 1.6 shows pictorially the two types of presses, *viz.* single-action and double-action, while the third sketch represents the inverted double-action variety.

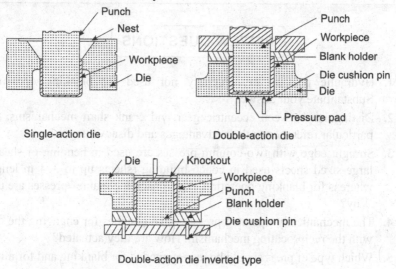

Fig. 1.6 Single-action; Double-action; Inverted Double-action Presses

SUMMARY

In this chapter, as a prelude to design of press tools, various types of commonly used presses are explained. Basically, press tools are classified into four types: (*a*) based on the type of power source, (*b*) based on the type of frame, (*c*) based on the method of actuation of slide and (*d*) based on the number of slides employed. While mechanical presses can achieve high production rates like 50 strokes/min, hydraulic presses can be employed for large strokes and for deep drawing operations. Hydraulic presses can generate nearly 10 times more force than mechanical presses and can also be used for large components.

Similarly, comparison of gap-frame presses with that of straight-side presses has been made. Gap-frame presses are intended for stamping functions involving blanking, piercing, coining, etc., whereas straight-side presses are intended for bending, shearing and blanking of large panels used in coach and bus bodies. It has a larger capacity, in addition to being resistant to misalignment, particularly in off-centre loads.

Comparison is also made on the classification based on the number of slides. Although the presses are classified into four categories, as per the committee of press builders (Joint Industry Conference), the presses can be classified into 18 categories.

REVIEW QUESTIONS

1. Hydraulic presses are generally not used for blanking operations. Substantiate your answer.
2. Distinguish between eccentric gear and crank shaft mechanisms, with particular reference to their advantages and disadvantages.
3. Straight edge with two-column presses are used in bending or shearing large-sized sheets used in coach building (sheets up to 3–4 m length); whereas for blanking operation, "open back inclinable" presses are used. Why?
4. The mechanically operated presses have clutches for engaging the drive with the reciprocating mechanism. How are they actuated?
5. Which type of press should be recommended for blanking and forming of large stainless steel bath tubs?

□□□

Sheet Metal Forming Processes

2.1 CLASSIFICATION

In this chapter, a brief outline of various sheet metal forming processes are described, with particular reference to press tool applications. Forming processes such as rolling and tube forming are not dealt here. Basic types of sheet metal forming using press tools are given as follows:

1. Processes related to Shearing or Cutting:

(*a*) Blanking   (*b*) Cut-off

(*c*) Parting   (*d*) Piercing

(*e*) Notching   (*f*) Lancing

(*g*) Trimming   (*h*) Shaving

(*i*) Fine Blanking

2. Processes related to Bending:

(*a*) Bending in 'V' die

(*b*) Bending in 'Wiping' die

(*c*) U-bending die

3. Processes related to Forming

4. Processes related to Drawing

Drawing and deep drawing: In the following paragraphs, disruption with figures are provided for each of the above processes. General formulae for calculation of forming forces are also given.

(*a*) **Blanking:** It is the cutting of the complete outline of a workpiece in a single press stroke. Since the scrap skeleton is produced, this process involves waste. However, this process is particularly suitable for fast and economical production. In order to achieve a saving in the material cost, suitable blank layout with the selection of stock width is done. This calls for several trial layouts. If A_b represents the area of blanks produced in one press stroke and A_s the strip area consumed by one press stoke or strip

Design of Jigs, Fixtures and Press Tools, First Edition. K. Venkataraman.
© K. Venkataraman 2015. Published by Athena Academic Ltd and John Wiley & Sons Ltd.

width times feed length, then the percentage of scrap in strip layout can be calculated as:

$$100 \left(1 - A_b/A_s\right)$$

Staggering is one technique used in producing circular blanks. Figure 2.1 shows the strip layout for such blanks, reducing the material loss to a minimum. With irregular-sized blanks, nesting or interlocking of blanks in the layout is done to maximize production. Figure 2.2 shows the nested layout and Fig. 2.5 shows a simple punch and die for carrying out blanking.

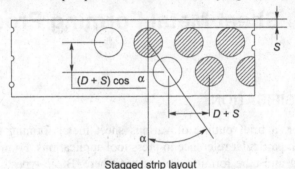

Stagged strip layout

Fig. 2.1 Zig-zag Layout

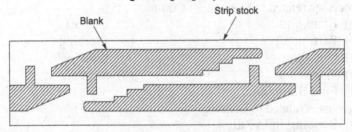

Nesting of irregular blanks in layout to save material

Fig. 2.2 Nested Layout

(*b*) **Cut-off:** In this operation, cutting is done along a line so as to produce no scrap. The cut-off line can take any profile. After the cutting-off operation is performed, the blank falls into the chute or into a conveyor. Figure 2.3 shows an example for cut-off.

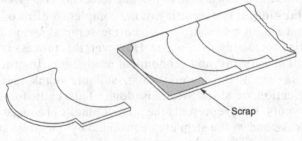

Cut-off

Fig. 2.3 Example for Cut-off

(c) **Parting:** If the adjacent blanks do not have mating adjacent surfaces, parting can be carried out. Figure 2.4 shows an example for parting.

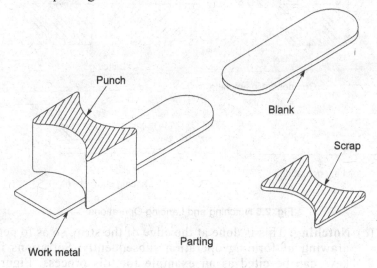

Fig. 2.4 Example for Parting

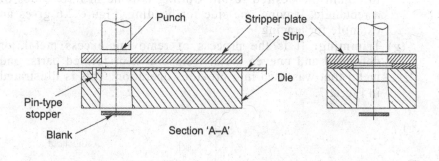

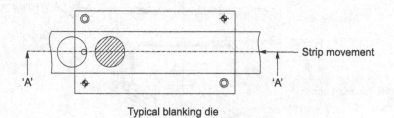

Typical blanking die

Fig. 2.5 Typical Blanking Operation

(d) **Tiercing:** .....

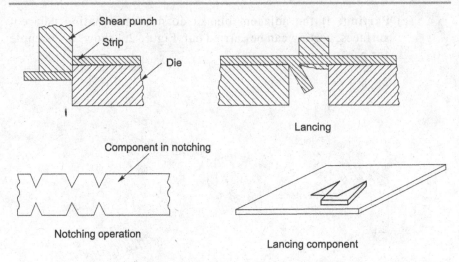

Fig. 2.6 Notching and Lancing Operations

(*e*) **Notching:** This is done at the edge of the strip, so as to perform drawing of forming operation subsequently. Serrations in the keys can be cited as an example for this process. Figure 2.6 gives an example for notching.

(*f*) **Lancing:** Both cutting and bending operations are performed to obtain the desired result. Cutting is done in three sides of a rectangle, leaving one side for bending. Figure 2.6 gives an example for lancing.

(*g*) **Trimming:** It is the process of removing excess metal, or deformed and uneven metal from drawn or formed parts, and metal that was used in a previous operation. This is illustrated in Fig. 2.7.

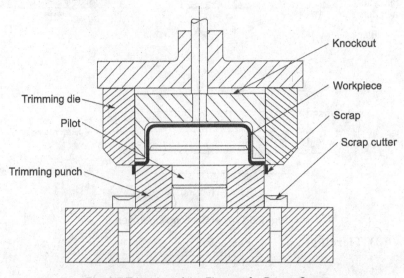

Fig. 2.7 Trimming of the Flange of a Drawn Cup

(*h*) **Shaving:** In this process, deformed, broken and burred edge, that was left in blanking, is removed. The scrap resembles the chips produced in machining. Two shaves are performed to get a better and straighter edge than a single shave. This process produces more wear on the die, since the punch to die clearance is closer than in blanking.

(*i*) **Fine blanking:** The basic differences between blanking and fire blasting are:

   (*i*) 'V'-shaped impingement ring restricts the flow of material away from the cutting edge.

   (*ii*) Triple-action presses are used: the first action for clamping, the second one for counter-pressure and the third one for blanking operation.

   (*iii*) Clearances between punch and die are very close (say, 2% of thickness of stock); hence the entire length of the stock is burnished, unlike in blanking, where burnishing is followed by fracture.

   (*iv*) Holes have a diameter of even upto 50% of stock thickness. Figures 2.8 and 2.9 explain the process sequence.

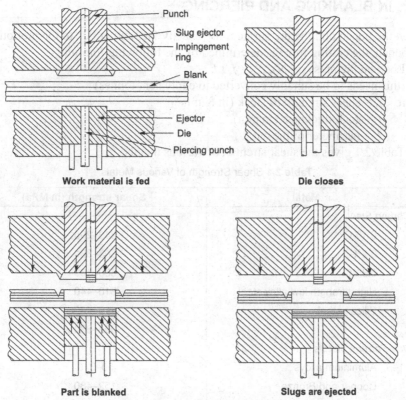

Work material is fed   **Die closes**

Part is blanked   **Slugs are ejected**

Fig. 2.8 Example for Fine Blanking

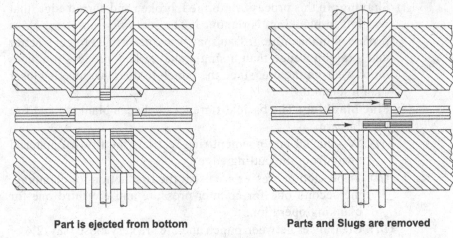

Part is ejected from bottom　　　　**Parts and Slugs are removed**

Fine edge blanking process (sequence)

Fig. 2.9 Example for Fine Blanking (Contd.)

2.2　CALCULATION OF FORCE REQUIREMENTS IN BLANKING AND PIERCING

When the punch and the die surfaces are exactly flat, and at 90° to the motion of the punch, the cutting force is determined by the following formula:

Load on the press (in N), $F_{sh} = f_s\, t\, L$

(this needs to be suitably converted to equivalent tonnes)

where　f_s = shear strength of stock (in N/mm^2)

t = stock thickness (in mm)

L = perimeter of the cut.

Table 2.1 gives the shear strength of various metals.

Table 2.1 Shear Strength of Various Metals

Metal	Shear strength (in MPa)
Carbon Steel	
0.1%C	240–295
0.2%C	303–380
0.3%C	360–460
High-strength low-alloy steel	310–440
Silicon steel	415–480
	395–830
Non-Ferrous Metals	
Aluminium alloys	50–310
Copper and Bronze	152–480
Titanium alloys	415–480

In addition to shearing force F_{sh}, stripping force F_{st} is also to be calculated to obtain the ejection force needed.

F_{st} is given by,

$$F_{st} = KA$$

where K = stripping constant, which depends on the thickness of the strip

A = area of the cut surface.

2.3   DIE CLEARANCES IN BLANKING AND PIERCING

Generally, clearances between die and punch are mentioned per side and they vary between 3 and 12.5% of the stock thickness of steel. Principles involved in deciding the clearances are:

(a) Plastic deformation and burnish depth are greater in thick material than in thin material and are greater in softer material. Figure 2.10 shows the mechanics of fracture that occurs while shearing the material and also shows various terminologies.

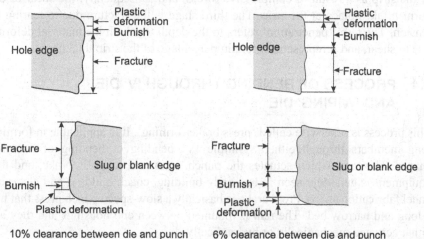

10% clearance between die and punch   6% clearance between die and punch

Fig. 2.10 Mechanics of Fracture

(b) Clearances vary directly with material thickness and inversely with ductility.

(c) The type of edge profiles of fractured blank or hole depends on the type of material as well as on the clearance. This means that for an edge of a fractured blank which has a small burnish, when compared to the fractured depth, large clearances of upto 17–21% may be selected. However, for straight edges having large burnish compared to the fractured depth, clearances of 1–2% of thickness of stock is selected. (Burnish is the depth of stock sheared, whereas fracture depth represents the depth of stock actually fractured due to excessive tensile stress.)

(*d*) Table 2.2 gives the percentage clearance recommended by a manufacturer for strip thicknesses of less than 3.18 mm.

Table 2.2 Recommended Clearances between Die and Punch

Material	Clearance per side percentage of stock thickness	
	Average (%)	Range
Aluminium alloys	2.25	1.7–3.4
Cold rolled steels (soft)	3.0	2.25– 4.5
Cold rolled steels (half-hard)	3.75	2.8–5.6

Depth of penetration

As explained earlier, the fracture of the strip under shear occurs when the tensile stresses due to penetration of the punch at the strip reaches a critical breaking strength of the material. In other words, the initial deformation is due to plasticity of the strip as a result of compressive forces, and subsequently, the shear or the burnish takes place at the strip. The third stage is the fracture due to tearing or tension. The term 'penetration' refers to the depth to which the material deforms due to shear, and it represents a certain percentage of the strip thickness.

2.4   PROCESS OF BENDING THROUGH 'V' DIE AND 'WIPING' DIE

This process is otherwise called 'press brake forming'. It is applicable in forming long members through either a simple 'V' bending or bending of complex shapes. The ram, which actuates the punch, is known as press brake, and this equipment is generally seen in bus body building, coach building and electrical panel fabrication shops. Press brake is basically a slow-speed punch press that has a long and narrow bed. The ram is mounted between end housings and they are either actuated mechanically or hydraulically.

The capacity of a press brake ranges between 70 KN and 20 MN, and is decided by the bending force
given by

$$F_{be} = \frac{KLt^2S}{W}$$

where L = length of bend (in mm)
t = thickness of stock (in mm)
K = die opening factor, varies between 1.2 for a die opening of 16 t
to 1.33 for a die opening of 8t
S = ultimate tensile strength (in N/ mm^2)
W = width of die opening (in mm).

The nose radius of the punch should not be less than 1t for bending mild steel, and it should be increased as the formability of the metal increases. In order to permit the punch to reach the bottom of the die, the die radius must be greater than the punch radius by an amount equal to the stock thickness. Earlier, a brief outline on the press brakes was given. However, the following points are also to be noted in regard to the selection of press brakes:

- Ram of the press advances rapidly just up to the stock
- The ram slows down just above work
- The ram proceeds slowly up to the bottom of the stroke
- The ram returns rapidly.

Another type of bending die is the 'wiping' die. Figure 2.11 shows the difference between an ordinary 'V' die and a 'wiping' die. In this, the pressure pad, which is loaded either by spring or by fluid cylinder, clamps the workpiece to the die before the punch makes a wiping action. In order to restrict the severity of the wiping action, punches have radii or chamfers.

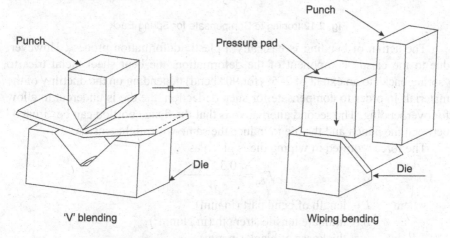

Fig. 2.11 V-Bending and Wiping Bending

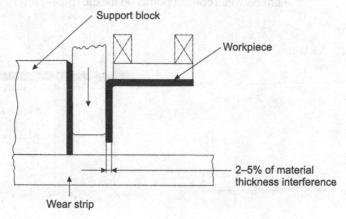

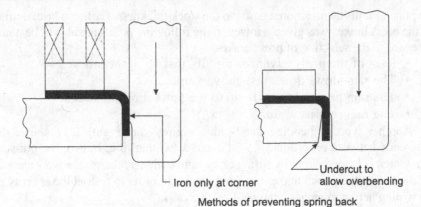

Methods of preventing spring back

Fig. 2.12 Ironing to Compensate for Spring-Back

The action of bending is basically a plastic deformation process. However, due to the elastic component of the deformation, the bent sheet metal tries to "spring back" to an extent of 2–5° (for 90° bend) depending on the ductility of the material. In order to compensate for such deflection, the die is undercut to allow for overbending. The second alternative is that the flange portion can be "ironed" between the punch and the die to induce the same with slight tension.

The force required in wiping die is given as

$$F_{be} = \frac{0.33\ LSt^2}{W}$$

where L = length of bend part (in mm)
S = ultimate tensile strength (in N/mm^2)
t = thickness of blank (in mm)
W = width between contact points on the die (please refer Fig. 2.13).

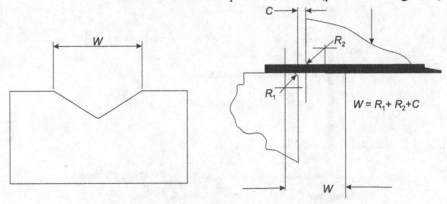

Fig. 2.13 Width of Die in 'V' Die and Wiping Die

Some of the terminologies in bending are explained in Fig. 2.14. For calculating the bend allowance, which is nothing but the length of the neutral axis of the strip, the following formula is applied:

Bend allowance, $$B = \frac{A \times 2\pi[IR + Kt]}{360}$$

where
A = bend angle

IR = inside radius of bend (in mm).

K = constant for neutral axis location (0.33 for $IR < 2t$ and 0.50 for $IR > 2t$).

t = metal thickness (in mm).

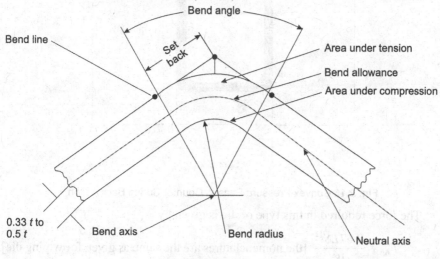

Fig. 2.14 Nomenclature in Bending

'U' bending or channel bending dies are illustrated in Fig. 2.15. Side clearance in this type of dies should be 10% more than the stock thickness.

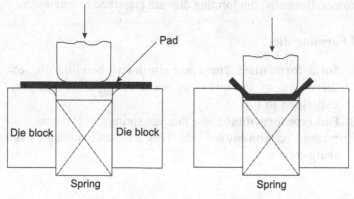

Fig. 2.15 'U' or Channel Die

'U' dies/Channel dies are equipped with pressure pads, as shown in Fig. 2.15. Spring back can be compensated by providing convex pressure pad as shown in Fig. 2.16.

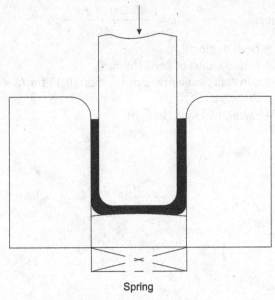

Spring

Fig. 2.16 Convex Pressure Pads to Counter Spring Back Action

The force required in this type of die is given by

$$F_{bc} = \frac{0.67 LSt^2}{W} \text{ [the nomenclatures are the same as given for wiping die]}$$

2.5 FORMING DIES

Forming operation is done along a curved axis rather than a straight axis. Due to the complexity of shapes, general formulae are not developed for determining forming forces. Basically, the forming dies are classified in hereunder.

Types of Forming dies

1. **Solid form dies:** These are similar to bending die, except that these are of complex shapes like pipeline clamps. This is explained in Fig. 2.17.
2. **Pad-type form dies:** These dies use springs, or pneumatic or hydraulic pressure pads on any one side. They are used in intricate shapes having sharp corners.

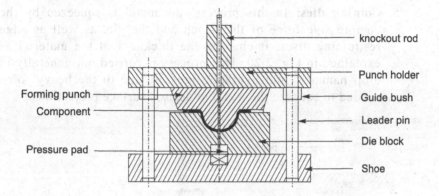

knockout rod

Punch holder

Forming punch

Guide bush

Component

Leader pin

Die block

Pressure pad

Shoe

Fig. 2.17 Solid Forming Die (Sectional Front View)

3. **Curling dies:** Curling is done at the edges of the drawn component. Many of the stainless steel utensils have curled edges. This is explained in Fig. 2.18.

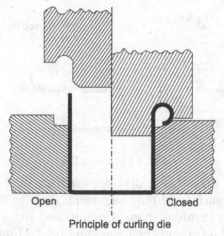

Open   Closed

Principle of curling die

Fig. 2.18 Curling Die

4. **Embossing dies:** Here there is no change in thickness of the metal, as the component takes the shape of the male punch. Here, one side has protrusions, whereas the other side has depressions. The same is explained in Fig. 2.19.

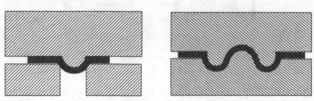

Fig. 2.19 Embossing Die

5. **Coining dies:** In this process, the metal is squeezed by the compressive force of the punch and the die, as well as edge restraining discs. It changes the thickness of the material as explained in Fig. 2.20. This process is carried out generally in drop hammers and hydraulic presses, due to the heavy force needed to squeeze the metal to desired surface projections.

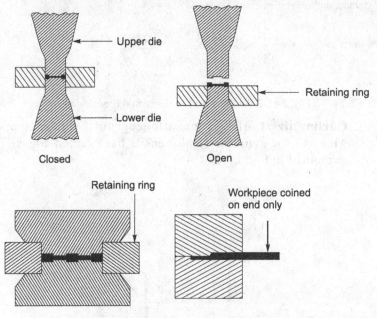

Fig. 2.20 Coining Die

6. **Bulging dies:** These dies use urethane or fluid medium at the internal surface of the component/ dish, so that the component takes the required shape. Complicated shapes are produced by this method. This is explained in Figs 2.21 and 2.22.

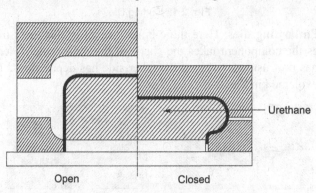

Fig. 2.21 Bulging Die with Urethane Medium

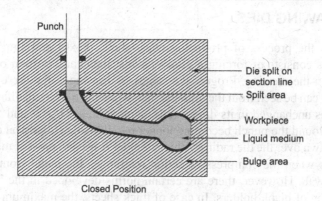

Fig. 2.22 Bulging Die with Liquid Medium

7. Assembly dies: These dies are employed in riveting or joining of two members. Figure 2.23 illustrates the process.

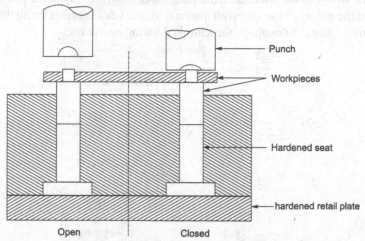

Fig. 2.23 Assembly Die for Riveting Operation

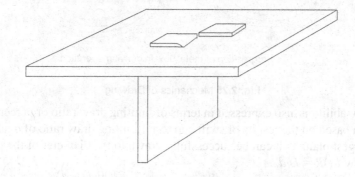

Fig. 2.24 Components Assembled in Assembly Die

2.6 DRAWING DIES

Drawing is the process of producing cups, shells, boxes and similar articles. The process consists of forcing a punch against the centre portion of the blank that rests on the die ring. Progressive stages in drawing a cup are explained in Fig. 2.25. It can be seen from the drawing that, while the flat circular blank remains more or less unchanged in its dimensions, annular areas '1', '2' and '3' get bent and wrap around the punch becoming longer parallel-side cylindrical elements as they are drawn over the die radius. A blank holder is used to prevent the formation of wrinkles when the compressive action rearranges the metal from the flange to the sidewall. However, there are certain norms developed in the industry for the provision of blank holders. In case of thick sheets, the maximum percentage reduction in blank diameter to cup diameter will be 25% for drawing without blank holders, and if blank holders are employed, the percentage reduction can go up to 50% for metals having good draw-ability characteristics. (Draw-ability means the ability of the material in the flange area to flow easily in the plane of the sheet and the ability of the side wall material to resist deformation in the thickness direction). Figure 2.26 explains the sample drawing operation.

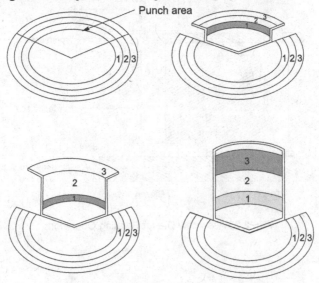

Progression of metal flow in drawing operation

Fig. 2.25 Mechanics of Drawing

Draw-ability is also expressed in terms of limiting draw ratio or percentage of reduction based on the results of swift cup test. Limiting draw ratio of diameter D of the largest blank that can be successfully drawn to the diameter of the punch d is given by $LDR = D/d$.

Percentage reduction would then be defined as: $\dfrac{100(D-d)}{D}$

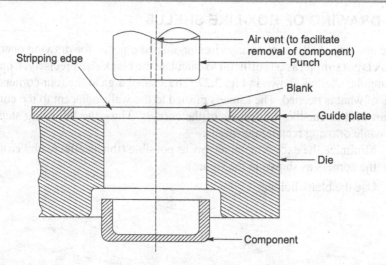

Fig. 2.26 Typical Drawing Operation

Theoretical maximum reduction in one draw is approximately 50%. However, for all practical purposes, if the percentage reduction exceeds 40%, a second draw is recommended. Similarly, if h/d exceeds 0.75, more than one reduction is necessary (where h is the height of the shell).

The drawing force F_d (max) is given by:

$$F_d(\text{max}) = \pi\, d\, t\, S\, (D/d - c)$$

where
- d = shell diameter
- t = thickness of the material
- S = yield strength of the material
- D = blank diameter
- C = constrict to cover friction (= 0.6–0.7 for ductile material).

The necessary theoretical calculations to give the blank diameter for a given shell diameter is given as follows:

$$D = \sqrt{d^2 + 4dh} \qquad \text{where } d/r > 20$$

$$= \sqrt{(d^2 + 4dh) - 0.5r} \qquad \text{where } d/r \text{ is between 15 and 20}$$

$$= \sqrt{(d^2 + 4dh) - r} \qquad \text{where } d/r \text{ is between 10 and 15}$$

$$= \sqrt{(d - 2r)^2 + 4d(h - r) + 2\pi r(d - 0.7r)}$$
$$\text{where } d/r < 10$$

where
- D = blank diameter
- d = shell OD
- h = shell height
- tr = corner radius of punch.

2.7   DRAWING OF BOX-LIKE SHELLS

Unlike drawing shells, cups or any circular-shaped objects, the drawing operation of a box is not simple, as it is difficult to calculate the blank size precisely. Consider a rectangular shell as shown in Fig. 2.27. The shaded area at the four corners is in excess of what is needed. The same is pushed to the walls adjacent to the corners, and part of it as ear-like extensions of the corners. Thus, the following steps are taken while drawing rectangular shells:

- Minimize the excess metal as low as possible (this is effected by cropping the corners as shown in the figure)
- Use the blank holders.

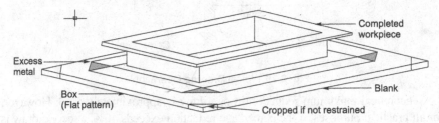

Fig. 2.27 Rectangular Component in Drawing

2.8   DIRECT AND REVERSE REDRAWING

In direct redrawing, there are two options: first one is the single-action redrawing and the second one is the double-action redrawing, and they are explained in Fig. 2.28. Double-action signifies the movement

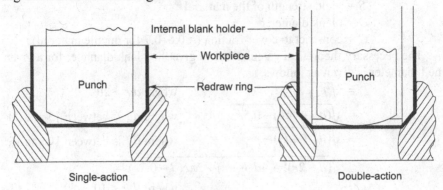

Fig. 2.28 Direct Redrawing

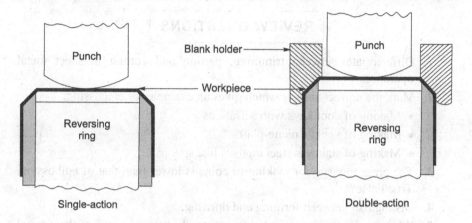

Fig. 2.29 Reverse Redrawing

of the internal blank holder initially and the punch subsequently; whereas in the case of single-action, there is no movement of internal blank holder. Only the punch moves downwards, making the redrawing possible. In the case of reverse redrawing, the redrawing is done as shown in Fig. 2.29. Here again, single and double-action is possible depending on the movement of the blank holder. Advantage of reverse redrawing is that the drawing and redrawing can be accomplished in one stroke of a triple-action hydraulic press or a double-action hydraulic press with a die cushion, eliminating the need of a second press. Greater reduction per draw is also feasible in reverse redrawing.

SUMMARY

In this chapter, various processes involved in sheet metal forming involving press tools are explained. They are basically classified as per the type of forming such as (a) cutting/shearing, (b) bending, (c) forming complex curved shapes and (d) drawing. Although all of the above deformation depends on the plasticity of ductile material, certain subtle differences in the type of stresses involved in each case exist. For example, blanking involves: initially plastic deformation, subsequently shearing, and finally fracture. In the case of bending, the deformation is initially elastic and then plastic. Once the load is removed. The elastic strain gets eliminated with a spring back action. As regards drawing, the compressive stresses in the annular space of the blank gets converted to tensile stresses once they take the shape of cylindrical shells. Bottom portion, which is in contact with the punch, remains unchanged due to the absence of any stresses. Forces needed to deform in each of the above cases are also discussed in this chapter.

REVIEW QUESTIONS

1. Differentiate between trimming, parting and cutting in sheet metal operation.
2. Mark the correct answer which gives an example for embossing:
   - Making of door keys with serrations
   - Making of vehicle name-plates
   - Making of stainless steel utensils like spoons.
3. The press tonnage for making of coins is lower than that of embossing. True/False?
4. Distinguish between forming and drawing.
5. What are the methods adopted to counter the "spring back" of sheet metal while forming?
6. What do you understand by the term "Bend Allowance"? Explain its significance in calculating the bending force.
7. Depth of penetration and the thickness of stock are one and the same. True/False? Substantiate your answer.
8. (a) Reverse redrawing is a technique adopted in lieu of regular redrawing, so as to increase the productivity of the drawing die. True/False?
   (b) Reverse redrawing is a technique which will reduce the work hardening effect on the component, when compared to regular redrawing. True/False?
9. Why is that for calculating the force due to bending, the ultimate tensile strength is taken into account, whereas in the case of evaluating the drawing force, the yield strength is considered?
10. Explain why the drawing of box with flange all round needs more trimming allowances than for drawing of a cylindrical cup having flange all round.

3

Introduction to Press Tools

3.1 STANDARD DIE SET

In the previous two chapters, a brief outline of the types of presses and the forming processes relating to the sheet or strip that are generally carried out are explained as a prelude to the topic of press tool design. A press tool is a device that enables any one of the operations such as blanking, piercing, bending, and drawing to be performed repeatedly in mass-scale in a press. In some other occasions, operations like blanking and bending, blanking and drawing or piercing and blanking can be carried out in tandem, calling for specialised press tools. Since the punch or a die are the two major components in a press tool, which are subjected to severe loads and hence wear and tear, they are quite often designed to be replaced, retaining the body of the press tool, *viz.* 'Die Set' as a standard device in a press tool.

A Die Set consists of the following components:

- *Die Shoe*, which holds the die by means of a set of screws and dowels. They are designed to withstand the severe impact loads transmitted to the die. They have flanges protruding on the two ends, to facilitate fastening the die set to the bolster plate of the press. (A bolster plate is a rugged and heavy member of the press over which the press tool is mounted. The plate is connected to the base frame of the press allowing for transmission of transient impact loads).
- *Punch holding plate*, which moves up and down the guiding pins and which carries the punch.
- The *shank* is a member in the die set to which the punch plate is fitted. This is also fastened to the ram of the press and is a transmitting member of the reciprocating motion down to the component.
- *Guiding pins* having bushes are vertical pins about which the punch holder plate and hence the punch, moves vertically. These are fitted to the die shoe. Guiding pins can be two, three or four in number, depending on the application. But, generally, two pins are located at the rear of the press tool to provide for easy movement of the strip.

Design of Jigs, Fixtures and Press Tools, First Edition. K. Venkataraman.
© K. Venkataraman 2015. Published by Athena Academic Ltd and John Wiley & Sons Ltd.

Figure 3.1 shows the outline of a die set for a two-pin configuration. One important parameter in the design and selection of die set is the "shut height". This is nothing but the vertical height of the die set, between the bottom of the die shoe and the punch holder, measured when the device is in closed position. In addition to two-pin, three-pin configurations and four-pin configurations are also deployed depending on the requirement. Two-pin configuration has a distinct advantage for strip flow into the die, as well as for exit. Four-pin configurations are used in large-sized and thicker components requiring precise alignment.

Figure 3.2 illustrates a typical standard Die Set having two-pin configuration. Tables 3.1 and 3.2 give the standard dimensions of some typical die sets, both round and rectangular, respectively.

Die sets are generally made up of cast steel, forged steel or rolled steel.

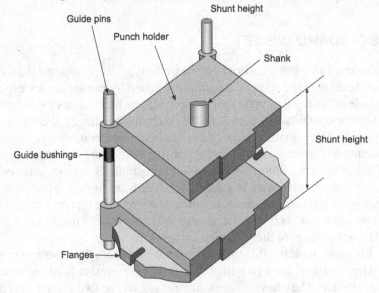

Basic components of standard die set

Fig. 3.1 Two-Pin Die Set

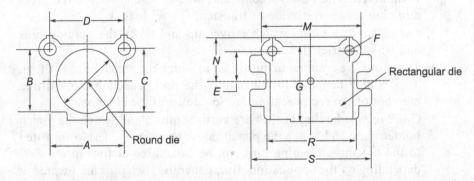

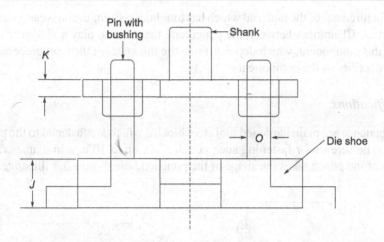

Fig. 3.2 Standard Die Set (both Rectangular and Round)

Table 3.1 Standard Dimensions of Some Typical Die Sets (Round)

A-Die Space	B-Die Space	Diameter of Round Dies	J-Die Holder	K-Punch Holder	C	D
75	75	75	31	25	84	81
125	125	125	50	45	136.5	100
250	250	250	56	45	266	256

Table 3.2 Standard Dimensions of Some Typical Die Sets (Rectangular)

A-Die Space	B-Die Space	J-Die Holder	K-Punch Holder	C	D	E	F	G	M	N	O	R	S
75	75	31	25	84	81	45	25	98.5	134	70	20	131	170
125	125	50	45	136.5	100	67.5	31	150	166	100	22	175	220
250	250	56	45	266	256	131	41	288	341	173.5	31	306	372
500	300	56	50	320	420	156	50	341	520	212	38	556	622

3.2 DESCRIPTION OF PRESS TOOLS

In addition to the die set, which is a standard device, the press tool has various other components. The various components in a press tool, their functions and their further classifications are given as follows.

3.2.1 The Blanking/Piercing/Bending/Drawing Punch

Major function

The basic function is to produce the desired quantity of components, whether it is blanking, bending or drawing, with the specified quality. As the punch is the main member, which imparts the required force of the press, the same needs to be manufactured out of the material which has toughness, strength and wear resistance properties. Clearances between the punch and the die also play a vital role in the life of this component, which depends upon the thickness of the workpiece as well as the ductility of the component

Classifications

Plain punches are plain hardened tool steel blocks, which are fastened to the punch plates. The screws for fastening such punches are up to 10 mm in diameter. The profile of the punch takes the shape of the punched hole. Figure 3.3 illustrates this type.

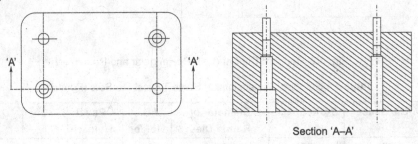

Section 'A–A'

Plain Punch constructed from solid block of Tool Steel

Fig. 3.3 Solid Punch–Fixing Arrangement

Pedestal punches are sometimes called Flanged punches because they are machined from a block of steel such that the punch has flanges all around. The flange portion is fastened to the punch holder. Figure 3.4 illustrates this type.

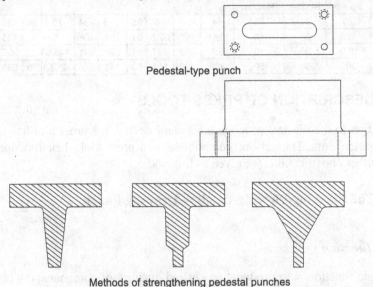

Pedestal-type punch

Methods of strengthening pedestal punches

Fig. 3.4 Pedestal Punches

Perforated punches are standard punches, which are usually less than 25 mm in diameter and are straight away screwed to the punch holder plate. The advantage of using these punches is that they can be removed and fitted without the removal of the punch holder plate. Figure 3.5 illustrates these types of punches.

Punch material depends on the material of the strip, its thickness and overall size. However, tool steels with high nickel are chosen because they retain hardness in high temperatures, and have strength, wear resistance and toughness to withstand crack propagation.

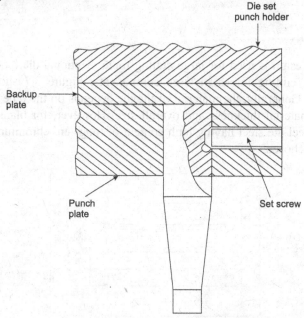

Fig. 3.5 Perforated Punch—Fixing Arrangement

Fig. 3.6 Methods of Mounting Perforator-type Punches

3.2.2 Blanking/Piercing/Bending/Drawing Die

Major function

These also play a vital role in the construction of press tools. These are designed according to the workpiece. Severity of forces are more predominant in blanking operations when compared to bending or drawing. Die material is generally of tool steel.

Classifications

Solid block design is used for large cutting force. Sectional die blocks are used for moderate cutting forces in large components. Figure 3.7 shows such an arrangement. Generally, the material of the die depends on the type of operation performed, material thickness and overall size. However, for blanking of steel strips, tool steel/die steel having high content of tungsten, chromium, vanadium and nickel is chosen.

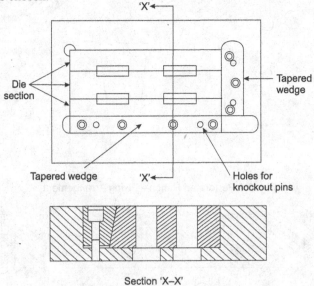

Section 'X–X'

Fig. 3.7 Die Section Nested in Pocket to Resist Heavy Cutting Forces

3.2.3 Stripper Plates

Major function

The strippers remove the strip from the blanking and piercing punches. The stripper is required since the material always tries to cling to the punch and there should be some provision to remove the same.

Classifications

Channel strippers are the common type in which a groove is milled in a plate for the strip to move. The plate is fixed to the die block through dowels and screws. The strip moves inside the grooved portion of the stripper. The groove depth is generally 1.5 times the strip thickness. The groove width is the same as that of the strip width plus any tolerance given for width variations. The overall thickness of the stripper plate depends on the socket screw head depth. Figure 3.8 provides the example for this type of stripper.

Spring strippers are otherwise called as pressure pads. The springs are selected based on the stripping force required. This is calculated based on the formula $F_{st} = KA$, where K is the stripping constant and A the area of the strip being stripped. In general, stripping pressure will be 5–20% of the cutting pressure. Figure 3.9 indicates the spring strippers.

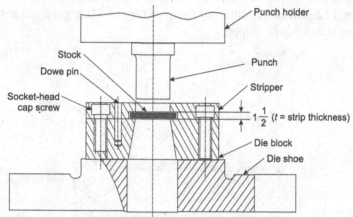

Fig. 3.8 Channel-type Stripper—Constructional Arrangement

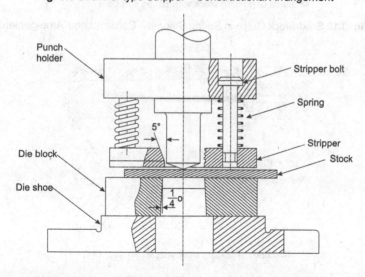

Fig. 3.9 Spring-loaded Stripper—Constructional Arrangement

3.2.4 Stock Guides

Major function

These are basically intended to guide the strip along the die for operations like blanking and piercing to be performed accurately in allotted locations.

Classifications

Guide rails are fitted to the die, and they do not come in contact with the spring strippers. It is illustrated in Fig. 3.10.

Button or spool-type strip guides are also provided. These have rotating motion while the strip moves. It is illustrated in Fig. 3.11.

Figure 3.12 provides additional examples for providing guidance to the strip when using channel-type strippers.

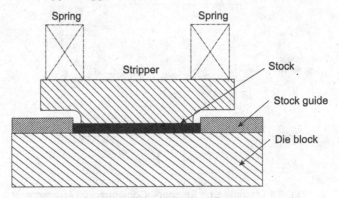

Methods of guiding stock with spring strippers

Fig. 3.10 Solid Stock Guide in Spring Stripper—Constructional Arrangement

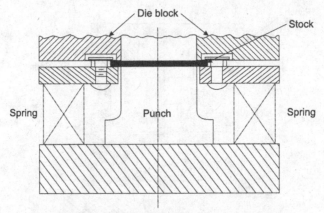

Methods of guiding stock with spring strippers

Fig. 3.11 Roller Stock Guide in Spring Stripper—Constructional Arrangement

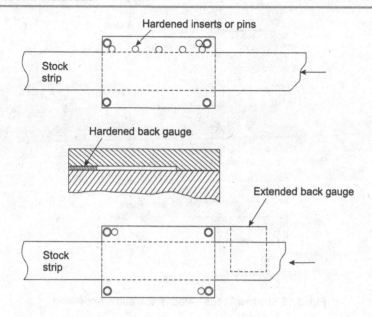

The use of back gauges in channel-type strippers

Fig. 3.12 Methods of Guiding Strips: Three Alternatives

3.2.5 Shedder

Major function

Due to the spring back of the blank, the same clings to the die. In other conditions, the lubricants make the blank stick to the punch, and this is called "slug pulling". To enable the blank to be removed from being "slug pulled", shedders are fixed to the punch with spring loads, so that the compressed spring ejects the blank once the operation is over.

Classifications

The first alternative is *mechanical mounting* of the punch under spring pressure.

The second alternative is by *pneumatic pressure* which is applied from the compressed air-line in the plant. Figure 3.13 illustrates both these *types*.

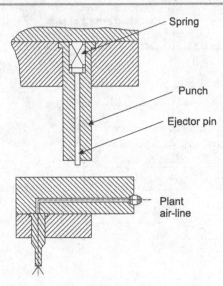

Fig. 3.13 Shedder Types—Mechanical and Air-operated

3.2.6 Stops

Major function

The function of stops is to restrict the motion of the strip, so that the desired operation such as piercing, blanking, cutting, parting, etc. can be carried out. In other words, the stopper is provided to facilitate the movement of the strip by a distance equal to 'Advance' ('advance' is the periodic distance moved by the strip, taking into account the blank size, so as to perform the desired task Ref. Fig. 3.15).

Classifications

Primary Stops are provided to perform the initial stopping of the strip when it is fed to the die set. Subsequent to the initial blanking or piercing or any other operation, this stop is redundant until the next coil is fed. Figure 3.14 explains this operation.

Trip Stops are provided to stop the strip during every 'advance' movement. This is spring actuated. Whenever the operator pushes the strip forward, the Pawl rises, and whenever the operator pulls the strip back, the same drops abutting the vertical face of the blanked hole. Figure 3.15 provides the illustration of such stops.

In addition to the above, simplest of all the stops, namely, pin or button die stops are generally provided to restrain the strip movement. The same is shown in Fig. 3.18. For every 'advance' movement, the operator pushes the strip over the pin stop, as the height of the pin protruding above the die surface will be generally lesser than the strip thickness.

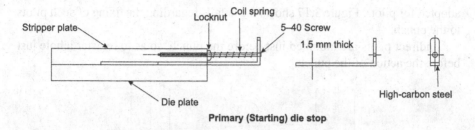

Primary (Starting) die stop

Fig. 3.14 Primary Stop for Starting Operation

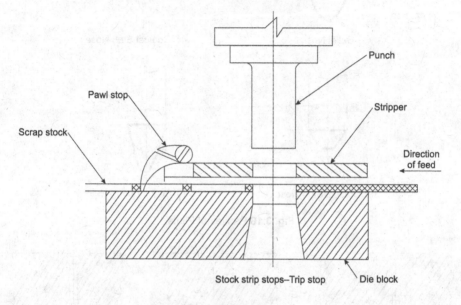

Fig. 3.15 Trip Stop

3.2.7 Pilots

Major function

The function of a pilot is to locate the strip or the component accurately on the required position, prior to a cutting operation. This positioning is known as registry action. Previously pierced holes are used for registry operation. If no holes are available in the component, additional holes can be made in the scrap area to perform the registry operation.

Classifications

Direct Pilots are mounted on the face of the punch and register on the hole pierced prior to the blanking operation. Figure 3.16 gives the profiles which are generally adopted for pilots. Figure 3.17 shows the details regarding the fixing of such pilots to the punch.

Indirect pilots are mounted just before the punch, so as to register a hole just before the action of the punch.

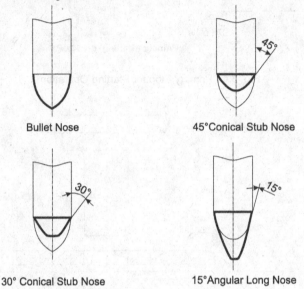

Bullet Nose   45°Conical Stub Nose

30° Conical Stub Nose   15°Angular Long Nose

Fig. 3.16 Profiles of Pilots

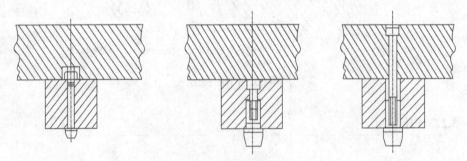

Fig. 3.17 Methods of Mounting Pilots to the Punches

Figure 3.18 shows the details of a blanking press tool with all the associated components. This will give an idea to the reader about the functions of the various components and the way they are assembled.

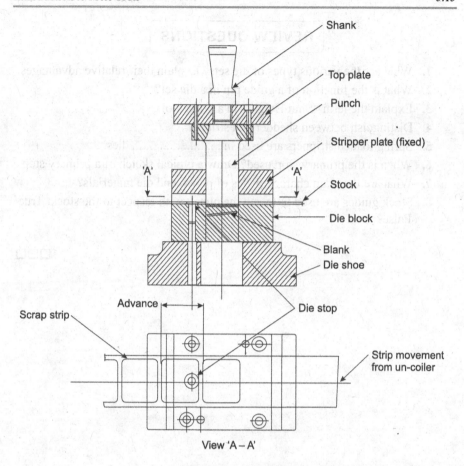

Fig. 3.18 Typical Blanking Die

SUMMARY

In the present chapter, an overview of a standard die set is provided with illustrations. Standard dimensions for such die is presented in the form of a table. Further detailed table providing such dimensions of the die set can be referred to in the Appendix.

In addition, various components of the press tool, such as punch, die, stops, guides, shedder, pilots, and strippers are explained with illustrations. Finally, a typical working drawing of a blanking press with all its associated components is shown as illustration. Components such as punch holder plate, die shoe and shank are also shown in the figure, so that the reader can have a comprehensive idea of the press tool.

REVIEW QUESTIONS

1. What are the various types of die sets? Explain their relative advantages.
2. What is the function of a guide pin in a die set?
3. Explain the term "Shut Height" in a press tool.
4. Distinguish between shedder and stripper.
5. Spring loaded strippers are used in .... dies.
6. When is the primary stop used? Draw a typical sketch of a primary stop.
7. What are the main characteristics of punch and die materials?
8. Stock guides are used to locate the punch with respect to the stock. True/False?

❏❏❏

Introduction to the Design of Blanking, Piercing, Progressive and Compound Dies

4.1 DESIGN OF BLANKING, PIERCING, PROGRESSIVE AND COMPOUND DIES

The processes of Blanking and Piercing were explained in Chapter 2. Both of them belong to the same category of material removal from an strip by shearing action. In mass production of blanks, the strip is fed from an un-coiler and is made to pass over the press tool for any given operation. The blanked or pressed strip is wound in a coiler placed opposite to the un-coiler or the exit side of the press tool. In certain other cases like production of washers or blanks that have smaller holes pressed in them, the operations, namely, piercing and subsequent blanking are performed in the same press tool. The piercing operation is performed in the first location and the blanking in the next location. Both these operations, however, are conducted in a single stroke of the press. Such dies are called progressive dies, due to the progressive operations being performed in the die. In Fig. 4.10, a progressive die is illustrated. In the same family of dies, a Compound die is also classified. In this die, both piercing and blanking operations are done in one station. All the above said dies are designed using the same principles which are listed in the next few paragraphs. In addition to die design principles, this chapter dwells on the method of determining the center of pressure for an unsymmetrically profiled blank, wherein the center of pressure does not lie at the axes of symmetry.

4.2 GUIDELINES FOR THE DESIGN OF PRESS TOOLS

Following are the guidelines followed for the design of press tools.

4.2.1 Scrap-Strip Layout

It may be appreciated that for the economical production of blanks, the utilization of the strip should be of high order, say at least 75%. This has been explained earlier in Chapter 2. Figure 4.1 illustrates a simple example of a scarp-strip layout.

Design of Jigs, Fixtures and Press Tools, First Edition. K. Venkataraman.
© K. Venkataraman 2015. Published by Athena Academic Ltd and John Wiley & Sons Ltd.

Generally, for strips whose thickness exceeds 0.75 mm, the following formulae are used:

t = thickness of strip

B = clearance between successive blanks or clearance between the edge of the strip and blank

$$= 1\frac{1}{4}t, \text{ for } C < 63 \text{ mm}$$

$$= 1\frac{1}{2}t, \text{ for } C \geq 63 \text{ mm Here,}$$

$C = L + B$-lead or advance of the die

where L = blank length

W = width of the strip

H = blank width or height.

For strips with thickness equal to or less than 0.625 mm, the above formulae are not to be used. Instead, Table 4.1 is to be used.

Table 4.1 Strips widths and dimension

Strip width, W (mm)	Dimension B (mm)
0 –75	1.25
75 –150	2.30
150 – 300	3.00
≥ 300	3.75

So far, the parameters for single-row single-pass layout have been explained (please refer Fig. 4.3 later in this chapter for single-row single-pass layout). In case the layout is decided to be double-row double-pass, as shown in Fig. 4.1, the clearance B will follow the following rule:

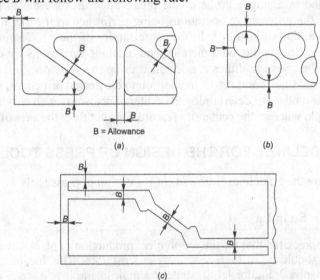

Fig. 4.1 Double-Row Double-Pass Layout

- Single-row double-pass, $B = 1\frac{1}{2}t$

- Double-row double-pass with curved lines, $B = 1\frac{1}{4}t$

- Double-row double-pass with straight and curved lines [as in Fig. 4.1 (c)], $B = 1\frac{1}{4}t$

The variants, namely, single-row double-pass or double-row double-pass are basically strip layout designed to improve the utilisation factor. Utilisation factor is the ratio of the blank area to the total area of the strip utilised to create a single blank. The same can be written in the form of equation:

$$\frac{a}{A} = \frac{LH}{CW}$$

It has already been explained that

$\quad L$ = length of blank
$\quad H$ = height or width of blank
$\quad C$ = advance or lead
$\quad W$ = width of strip

Generally, the utilisation factor is aimed at 70 – 75%.

4.2.2 Design of Die Blank

By the rule of thumb, for die blocks made up of tool steel for blanking materials ranging from low-carbon steels up to stainless steels, the thickness can be taken as given in Table 4.2.

Table 4.2 Perimeter of the blank and die thickness

Sl. no.	Perimeter of the Blank (mm)	Die thickness (mm)
1	0–75	20
2	75–100	25
3	100–175	31
4	≥ 175	38

As regards the top view of the die, *i.e.* its length and width, the same are decided as follows:

(*i*) Length of the die = Length of the blank (L)
$\qquad$ + twice the marginal clearance ($2 \times B$)
$\qquad$ + twice the clearance between the strip and the screws
$\qquad$ + twice the diameter of the set-screw or dowel
$\qquad$ + twice the margin from the set-screw till the edge of the die ($2 \times 10 = 20$ mm)

(*ii*) Width of the die = Width of the strip (W)
$\qquad$ + twice the marginal clearance ($2 \times B$)
$\qquad$ + twice the clearance between the strip and the screws

+ twice the diameter of the sub-screw/dowel (2×12)

+ twice the margin from the set-screw till the edge of the die ($2 \times 10 = 20$ mm)

Thus, for a blank opening of 60 mm × 60 mm, the size of die shall be as follows:

Length of the die = $\{(60 + 2.5) + 2 \times 10 + 20 + 20\} \cong 122.5$ mm (say 122 mm)

Width of the die = $\{(60 + 2.5) + 2 \times 10 + 20 + 20\} \cong 122.5$ mm (say 122 mm)

(assuming that set-screws and dowels of 10 mm are chosen and that the thickness of the strip is 1 mm.) With reference to die opening, the size should be the same as that of the blank. It is ground for the blank size up to a depth of the thickness of the strip. Subsequently, a taper angle of 1.5 degrees is provided to allow for the blank to drop without jamming.

4.2.3 Punch Design

Although the die opening is exactly the same as that of the blank size, the punch sizes will be smaller than the die sizes by twice the clearance assumed per side. If a clearance of 3% of stock thickness is assumed per side, then the cross-sectional size of the punch will be = 0.94 (length of die) × 0.94 (width of die)

(**Note:** In case of rectangular punches).

In the case of the length of the piercing punch, the same can be calculated using the following formula:

$$L = \frac{\Pi d \{E \, d\}^{1/2}}{8\{S_s \, t\}}$$

L = length of punch

S_s = unit shear stress on the stock (in MPa)

E = modulus of elasticity

t = thickness of the stock (in mm)

d = diameter of the punched hole (in mm).

This formula is applicable if $d/t \geq 1.1$.

In the case of rectangular blanks, the term Πd will be substituted by the perimeter of the blank, viz. $L \times H$ and diameter 'd' by 'L'.

However, the length of the punch is generally assumed to be the blank length, or as the case may be for rigidity. Let us assume that the punch length is 60 mm and it is held against a hardened backup plate of 60 mm by a punch plate of 20 mm. The whole assembly is screwed and dowelled to the upper shoe or the top bolster plate. By the rule of thumb, the upper and bottom bolster plates are assumed to be (1.25 × thickness of die plates) and (1.75 × thickness of die), respectively. Refer to Fig. 4.2.

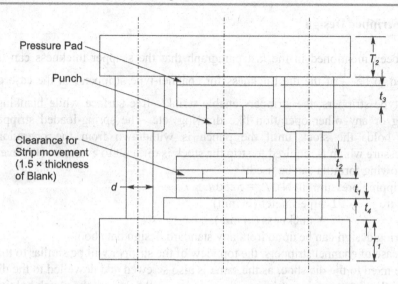

Nomenclature

t_1 : Thickness of Die Plate

t_2 : Thickness of Stripper Plate

t_3 : Thickness of Punch Holder Plate

t_4 : Thickness of Die Shoe

T_1 : Thickness of Bottom Bolster Plate

T_2 : Thickness of Top Bolster Plate

d : Diameter of Guide Pin

F_{sh} : Shearing Force (in tonnes)

Rule of Thumb for Design

t_1 : Thickness of Die Plate (in cm) = $3\sqrt{F_{sh}}$

t_2 : Thickness of stripper Plate $n = 0.5t$

t_3 : Thickness of Punch Holder Plate = $0.5t$

t_4 : Thickness of Die Shoe = t_1

T_1 : Thickness of Bottom Bolster Plate = $1.75\,t_1$

T_2 : Thickness of Top Bolster Plate = $1.25t$

d : Dimater of Guide Pin = t_1

Fig. 4.2 Method of Calculating the Die Thickness—Rule of Thumb

As per the rule of thumb, the stripper plate thickness will be half of the die plate thickness and for channel-type strippers, the channel clearance will be $1\frac{1}{2}t$. .

Thus, the total shut height for a 25 mm die plate will be:

Shut height = {die plate thickness + die shoe thickness

$$+ \left(\text{Stripper Plate thickness} + 1\frac{1}{2} \right)$$

+ top bolster plate thickness (1.25 × die plate thickness)

+ bottom bolster plate thickness (1.75 × die plate thickness)

+ (punch holder plate thickness + backup pressure pad thickness) + clearance of 10 mm}

= 25 + 25 + (12.5 + 1.5) + 31 + 45 + 18 + 10

(assuming 1 mm as the thickness of blank)

$\cong$ 168 mm = 170 mm (say)

4.2.4 Stripper Design

It has been mentioned in the last paragraph that the stripper thickness can be assumed to be $\frac{1}{2}$ of the die thickness for channel-type strippers. In the case of spring-type strippers, the strippers enable wrinkle-free surface while blanking, piercing or any other operation like drawing, etc. The spring-loaded stripper further holds the stock until the punch is withdrawn from the operation. The pressure which is required to strip the stock is difficult to evaluate. However, the following formula can be used:

   Stripping pressure (in N), $P_s = 5250\, L \times T$

   where L = perimeter (in mm)

                T = thickness (in mm).

   Spring design can be done from any standard design data book.

   In case of channel strippers, the top view of the stripper will be similar to that of a die fixed to the die shoe, as the same is also screwed and dowelled to the die plate. In the case of spring strippers, the top view will be to suit the punch holding plate as well as the diameter of the spring holding screws and the spring diameter.

4.3   DESIGN OF PROGRESSIVE DIES

It has been explained in the first introductory paragraph regarding progressive dies. This is an extension of a blanking or piercing die. Here, two or more shearing/cutting operations are performed in the same die.

   These operations are performed in successive stations, each having either a piercing or a blanking punch. However, all these punches move in unison as they are fastened to the same top bolster plate, which moves up and down due to each stroke of the press. Thus, a progressive die is an innovation towards increased productivity. In the following paragraphs, step-by-step procedure for the design of a progressive die is explained.

4.3.1 Scrap-strip Layout

The first step is to design the strip layout for a given component. The component involves two sets of piercing operations and a blanking operation; a single-row strip is sufficient for such operations. A three-station will be sufficient. Figure 4.3 illustrates the component and the strip layout, taking into account the clearances between each blank. The thickness of the blank is assumed to be 3 mm and the size of the blank is 60 mm^2.

$$\text{Clearance, } B \ = \ 1.25t = 3.75 \text{ mm}$$
$$\text{Width of the strip, } W \ = \ 60 + 2B = 67.50 \text{ mm (say 68 mm).}$$
$$\text{The Lead or Advance, } C \ = \ 60 + B = 63.75 \text{ mm (say 64 mm).}$$

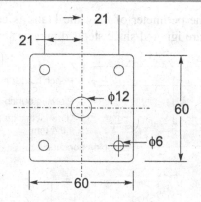

Component 3 mm thick

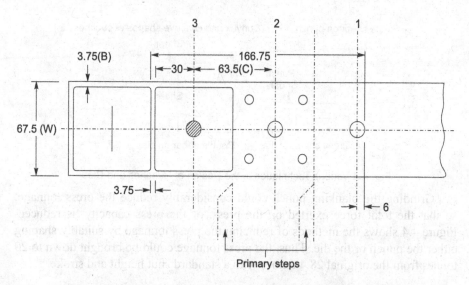

Fig. 4.3 Strip Layout in a Progressive Die showing the Component

4.3.2 Press Tonnage

The total force of the press is calculated as follows assuming shear strength as 395 N/mm² for cold rolled steel:

P = 395 × 40 = Perimeter of the cut length of the blank × 3t = 28.4 tonnes (approx.):

It may be noted that the perimeter of only the blank is considered, as the perimeter of pierced holes are ignored since stepped punches are used as shown in Fig. 4.4.

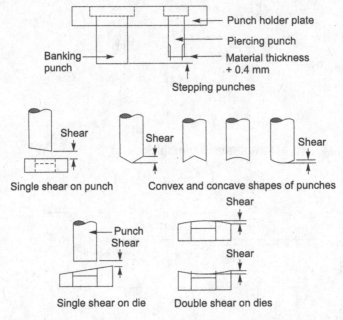

Fig. 4.4 Various Techniques in the Design of Punches and Dies

Grinding the blanking punch could considerably reduce the press tonnage, so that the total force exerted on the press, or the press capacity, is reduced. Figure 4.4 shows the methods of reducing the press tonnage by suitably shaping either the punch or the die. Thus, the press tonnage could be brought down to 20 tonnes from the original 28.4 tonnes with a standard shut height and stroke.

4.3.3 Design of Die Plate

Perimeter of the blank sheared out will be 240 mm. For such a perimeter, the die thickness will be 38 mm. As regards the overall length and width of the die, the following calculations show the procedure.

Width of the die

Width of the die is given by: Blank size + twice the Clearance on either side
$$60 + 62 = \textbf{122 mm}$$

Note: The clearances are obtained taking into consideration the size of the dowel pins or screws to be fitted to the die all round for fastening the same to the die shoe and the bolster plate. In this case, 31 mm clearance is assumed all round the strip, so as to allow for fitting the screws having diameters of 8– 10 mm.

Length of the die

Length of die is given by: $2 \times 64(C) + 30 + 6 + 2 \times 31$(clearance on either side) = **226 mm.** Hence, the size of the die plate will be **122 × 226 × 38.**

Figure 4.5 illustrates the die plate designed for a component chosen.

The die shoe size is given as:

$$(122 + 50) \times (226 + 50) \times (1 \times 38) = 172 \times 276 \times 38$$

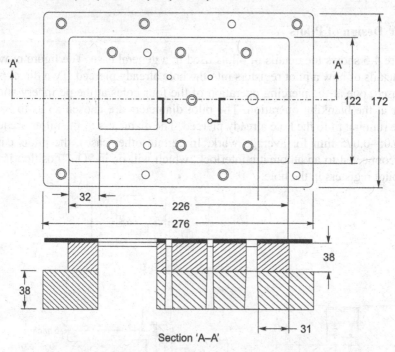

Fig. 4.5 Sectional View of a Die and a Die Shoe

The bolster plate at the bottom will be 1.75 times the thickness of the die plate designed and the size will be **310 × 405 × 66.5,** taking into account due allowances over and above that of the die shoe length and width. Adequate allowances are also provided for fastening the press tool to the table of the press.

4.3.4 Design of Punch

The blank size is 60 mm × 60 mm. Hence, the die size will be the same. However, the punch size will generally be lesser than the die size by an amount equal to 10% of the stock thickness. In other words, clearance provided per side will be 5% of thickness.

Punch size: $60 - 10\% \times 3 = 59.7$ mm

The height of the blanking punch can be the same as that of its lateral dimensions, *i.e.* 59–60 mm. However, in the case of piercing punches, the size of the punch will match with the size of the hole. The die size will be larger than the punch size by an amount equal to twice the clearance.

4.3.5 Design of Strippers

Channel-type strippers are used, as the same is enough for the given applications. The stripper thickness will be half of the die plate thickness, *i.e.* 19 mm. The channel depth will be 1.5 times the thickness of the stock, *i.e.* 4.5 mm. Stripper is fastened to the die by dowels and screws. Figure 4.9 illustrates the sectional view of the stripper.

4.3.6 Design of Pilots

Figure 4.6 shows the details of pilots used in a general case. The figure provides the details of how a pilot registers into the hole already pierced. Two direct pilots are used, one at the piercing operation of the four holes at the periphery and the other at the blanking operation. The pilot diameters are chosen so as to have a loose running fit to the hole already pierced. The diameters of the pilots are lesser by 0.05–0.075 mm for average work. In certain other cases, the pilots can be interconnected to an automatic interlock, which will be in "On" position if only the pilot registers in the hole.

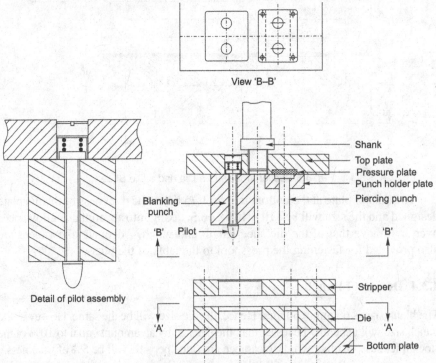

Fig. 4.6 General Design of Pilots and their Fastening Technique

4.3.7 Design of Stoppers

Two kinds of stoppers are used in this example. One is the primary stop, which is located in two locations to restrain the strip at the beginning. That is just before the second station, so as to perform the first operation of piercing the central hole. The second primary stop is located just before the third station, *i.e.* for performing the second operation of piercing four holes.

Once these two operations are performed, a simple pin stop or a trip stop (automatic stop) can be located just at the end of the third station. This stop will enable stoppage of the end of the strip at the beginning, as well as act as a stop even after the continuous operation starts. These are illustrated in Fig. 4.7.

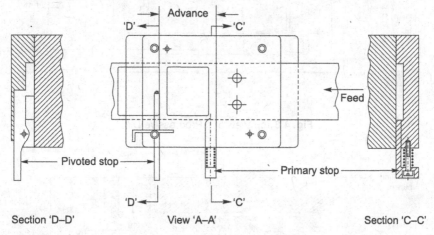

Fig. 4.7 Design of Stoppers Primary and Pivoted Stops in a Progressive Die

4.3.8 Shut Height of Die Set

The shut height of the die set is given as:
Top bolster thickness + punch holder and backing plate thickness put together
+ clearance of 10 mm between the bottom of the punch holding plate and the stripper plate
+ stripper plate thickness and the clearance for strip movement
+ die plate thickness
+ die shoe thickness
+ bottom bolster thickness
= $1.25 \times 38 + 25 + 10 + (19 + 4.5) + 38 + 38 + 1.75 \times 38$
= 249 mm or **250 mm**

Since the punch height has been chosen as 60 mm, the stroke of the press can be 50 mm. Thus, the press chosen shall have the following parameters:

- **Press Tonnage** : **20 tonnes**
- **Shut Height** : **250 mm**
- **Stroke** : **50 mm**

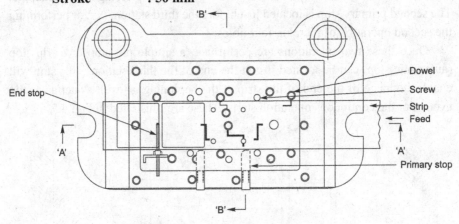

Fig. 4.8 Top View of a Progressive Die

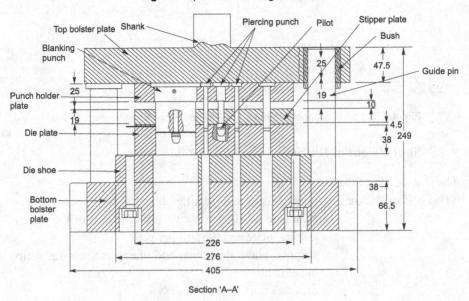

Section 'A–A'

Fig. 4.9 Sectional View of a Progressive Die

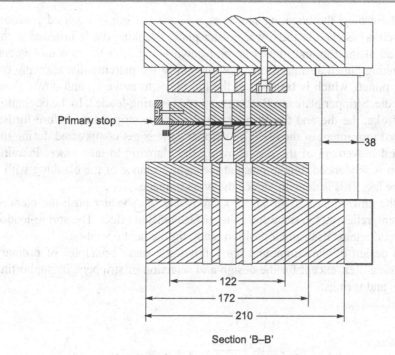

Primary stop

122
172
210

Section 'B–B'

Fig. 4.10 Sectional View of a Progressive Die

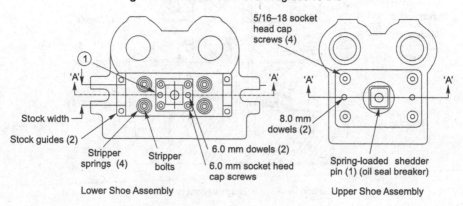

5/16–18 socket
head cap
screws (4)

①

'A'   'A'   'A'   'A'

Stock width

Stock guides (2)

8.0 mm
dowels (2)

Stripper
springs (4)

Stripper
bolts

6.0 mm dowels (2)

6.0 mm socket heed
cap screws

Spring-loaded shedder
pin (1) (oil seal breaker)

Lower Shoe Assembly

Upper Shoe Assembly

Fig. 4.11 Top View of Compound Die

4.4   COMPOUND DIE

These types of dies are explained in Figs 4.11 and 4.12. Both blanking and piercing operations are performed in a single stroke in a single station. Thus, such dies produce accurate components in terms of circularity and flatness. In addition, productivity of such dies is higher than the progressive dies.

In these dies, the blanking punch as well the die are in inverted position. The punch is fastened to the lower bolster plate. Blanking die is fastened to the punch pad at the top and moves up and down along with the ram movement. The blanking punch also performs the function of the piercing die, allowing the piercing punch, which is fastened to the top plate, to move up and down along with the die. Stripper plate is a floating type and is spring-loaded. In the beginning of the stroke, the die and the stripper plate grips the strip firmly before further downward movement of the die. The stripper springs get compressed during the downward movement of the die, allowing the blanking to take place. Piercing operation is sequenced to take place at the same instance of the blanking with a little time lag. This is done to reduce the press tonnage.

At the upward stroke, the knockout rods and the shedder push the blanked component, relieving the same from the surface tension effect. The spring-loaded stripper pad retains its original position ready for the next operation.

The design of the compound die follows the same principles of ordinary and piercing dies, except for the design and selection of strippers, its supporting elements and springs.

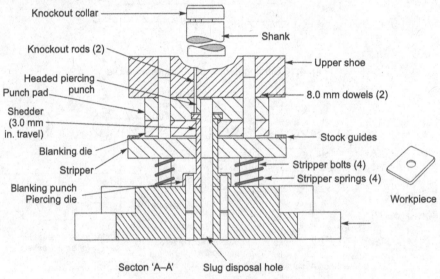

Fig. 4.12 Sectional View of a Compound Die

4.5 CALCULATION OF CENTRE OF PRESSURE IN UNSYMMETRICALLY PROFILED COMPONENTS

If a blanking component is not symmetrical about its axes, it is imperative that the force due to the press movement coincides with that of the center of pressure. Centre of pressure is defined as the imaginary point in the blank at which the resultant force due to shearing of the entire profile is suppose to act. If this is not ensured, then undesirable buckling loads may result. The same can cause quicker wear and tear of the punch and the die, failure of parts such as dowel pins and screws, and other maintenance problems. Therefore, the center of pressure is to be calculated precisely, and the press tool is to be located with respect to the press ram accordingly. Procedure for the calculation of the center of pressure is given in the Fig. 4.13 and is explained as follows:

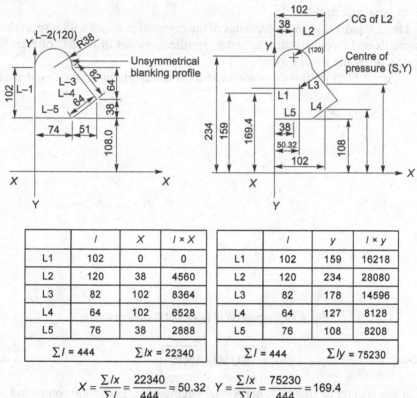

	l	x	$l \times x$		l	y	$l \times y$
L1	102	0	0	L1	102	159	16218
L2	120	38	4560	L2	120	234	28080
L3	82	102	8364	L3	82	178	14596
L4	64	102	6528	L4	64	127	8128
L5	76	38	2888	L5	76	108	8208
	$\Sigma l = 444$		$\Sigma lx = 22340$		$\Sigma l = 444$		$\Sigma ly = 75230$

$$X = \frac{\Sigma lx}{\Sigma l} = \frac{22340}{444} = 50.32 \quad Y = \frac{\Sigma lx}{\Sigma l} = \frac{75230}{444} = 169.4$$

All dimensions are in mm

Fig. 4.13 Calculation of Centre of Pressure

(*i*) Initially divide the profile into a number of segments, such as semi-circular curves, straight edges, circular edges, etc.

(*ii*) Assume imaginary X–Y axes adjoining the profile.

(*iii*) Multiply the length of each segment with the co-ordinate distances of the CG of each line segment.

(*iv*) Divide the sum arrived as per the procedure explained above in step (*iii*) with the sum of the lengths of all the line segments. The net result will be either the distance in x-axis of the center of pressure or the distance of center of pressure in y-axis.

The same can be expressed as follows:

$$X = (\Sigma x_n l_n) \Sigma l_n$$
$$Y = (\Sigma y_n l_n) \Sigma l_n$$

where $n = 1, 2, 3, \ldots$ and depends on the number of line segments assumed in the unsymmetric profile.

Here, X and Y are the co-ordinates of the center of pressure with respect to the assumed x and y co-ordinates. x_n and y_n are the distances of the CG of each line segment along the x and y axes.

Figure 4.14 illustrates the procedure for the determination of the CG of an arc.

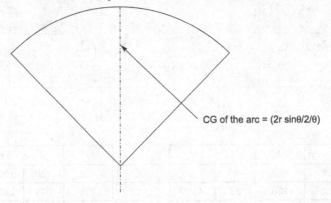

CG of the arc = (2r sinθ/2/θ)

Fig. 4.14 Calculation of Centre of Gravity of Arcs

SUMMARY

In this chapter on design of blanking, piercing, progressive and compound dies, a detailed account of the design of die plate thickness along with the design of other members of the press tool, *viz.*, die shoe, top and bottom bolster plates, stripper plate, primary and end stops and pilots are explained with sketches. The detailed design of a progressive die has been chosen as a case study. For a rough calculation of the die plate and other associated plate thicknesses, a rule of thumb formula with a typical sketch has also been provided, although the same is not meant for detailed design and analysis of press tool. Advantages of compound die and the constructional details are explained. Finally, a specimen calculation of an unsymmetrical profile for determining the centre of pressure is shown.

REVIEW QUESTIONS

1. Distinguish between single- and double-pass strip layouts.
2. Overall size of a die plate made of tool steel in a progressive die is decided based on the .... of the blank or on the .... force.
3. The capacity of the press is the same as that of the total shearing force assumed to act on the blank. True/False? Substantiate your answer.
4. The length of the piercing as well as blanking punch in a progressive die remains the same. True/False? Substantiate your answer.
5. If the shearing force of the punch does not act at the centre of pressure for an odd-shaped profile, one or more of the following will result:
   - Rapid wear and tear of the die and punch
   - Crack and subsequent fracture in either the punch or the die or both
   - Failure of the hold-down bolts at the press table
   - Failure of the components like punch holding plate/bolster plate
   - All of the above.

   Mark the right answer.
6. Why is tool steel/ die steel chosen for die material? What is the most important characteristic of the die material?

ANNEXURE TO DESIGN OF PROGRESSIVE DIES (CHAPTER 4)

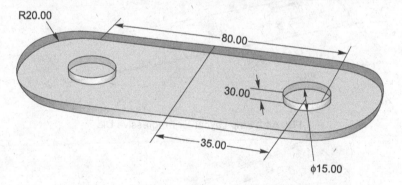

Fig. A4.1 Component Chosen for Progressive Die

PROGRESSIVE DIE DESIGN

The component shown in Fig. A4.1 is chosen for the design of a progressive die. It is assumed that the material is a medium carbon steel having an ultimate shear strength of 0.04 tonnes/mm². It is proposed to carry out the piercing of two holes, each having a diameter of 15 mm. Subsequently, blanking is done in the progressive die. The strip thickness is assumed as 3 mm.

Based on the said data, the total force required for blanking will be in the order of 34 tonnes (285 × 3 × 0.04). The piercing punches are designed to have a longer length than the blanking punch by 3 mm, thus enabling the piercing to be carried out with a lead time. Further, by having the blanking punch ground at its cutting edge in a sloping manner to the horizontal axis, the requirement of blanking force can be reduced to two-third of the calculated force, *i.e.* 22.6 tonnes or the press tonnage can be rounded to 25 tonnes. The press can have a stroke of 50 mm. The strip layout is done subsequently and the same is shown in Fig. A4.2. This enables the proper design of the die, including its thickness, which can be of 37 mm for a peripheral length (of blanking) of 286 mm. The die is shown in Fig. A4.3. The other components are designed based on the die design, and the progressive die thus designed is shown in Fig. A4.4 in the exploded view, which shows all the components involved, *viz.* (*a*) the Die, (*b*) Stripper plate, (*c*) Punch holding plate, (*d*) the Blanking and the Piercing punches, (*e*) Primary and end stops, (*f*) the Guiding pins, (*g*) the Guiding bushes, (*h*) Die shoe, (*i*) the Shank and (*j*) the Dowels and Screws for fixing the die with the die shoe as well as fixing of the punches with the punch holding plate. Figure A4.5 shows the wire frame model of the die. This facilitates the reader to have a view of the components hidden in the overall assembly.

Figures A4.6 – A4.10 show detailed models of various modules/components that make up for the complete assembly of a progressive die. Figure A4.11 illustrate a perspective view of a complete model of a progressive die.

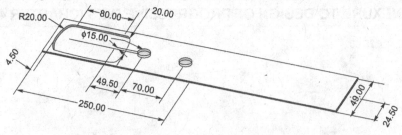

Fig. A4.2 Strip Layout for Progressive Die

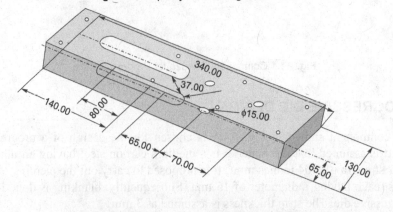

Fig. A4.3 Progressive Die with Critical Dimensions

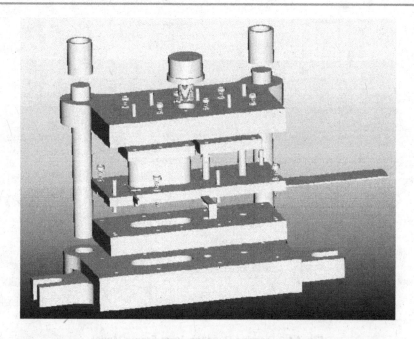

Fig. A4.4 Exploded View of the Progressive Die showing the Strip and Other Components

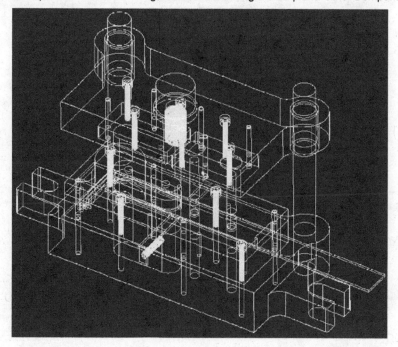

Fig. A4.5 Wire Frame Model of a Progressive Die showing the Hidden Lines

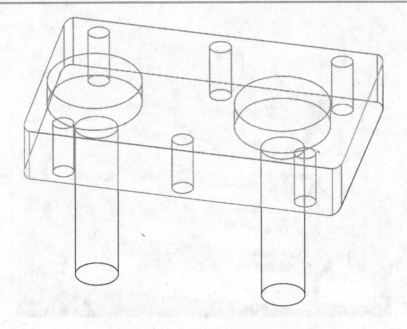

Fig. A4.6 Piercing Punches–Wire Frame Model

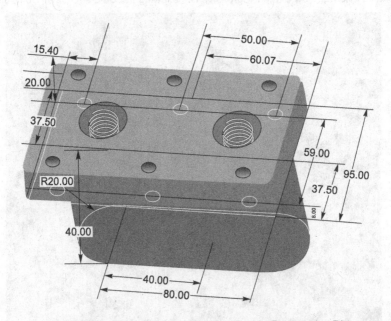

Fig. A4.7 Solid Model of the Blanking Punch–Progressive Die

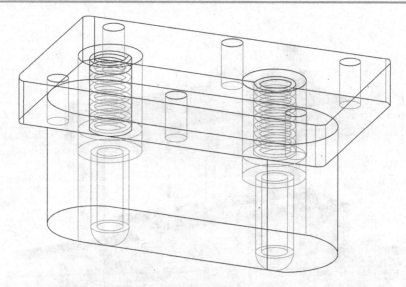

Fig. A4.8 Wire Frame Model of Blanking Punch showing the Pilots–Progressive Die

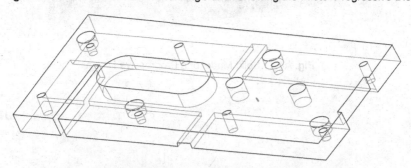

Fig. A4.9 Wire Frame Model of the Stripper Plate

Fig. A4.10 Solid Model of the Bolster Plate–Progressive Die

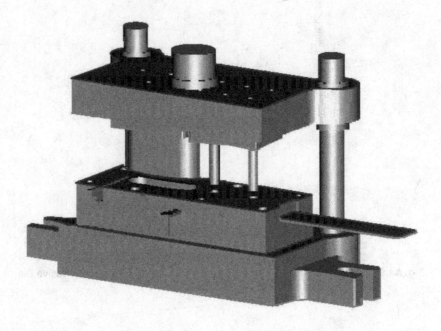

Fig. A4.11 Solid Model of a Progressive Die

❑❑❑

Bending , Drawing and Forming Dies

5.1   INTRODUCTION

In the previous chapter, a treatise on blanking, piercing and progressive dies is furnished, which deals exclusively with the process of shearing a strip. The design of such dies involves determination of the shearing force, which in turn depends on the material and the total shearing area. Once the shearing force is determined, the press capacity, the shut height and the stroke are selected, which are generally obtained from the standard information provided by the press manufacturer.

The die thickness is selected based on the peripheral length of the sheared portion. Thickness of the rest of the items such as strippers, die shoe, punch holding plate, etc. are chosen based on the die thickness. The top view of the die or the overall size of the die depends upon the blank size or the strip layout chosen.

However, in the case of bending, drawing and forming dies, although the forming forces are calculated as per the formula and procedure, selection of the die thickness depends mostly on the depth of bend, or the depth of draw. In many occasions, such as combination die, where both blanking and drawing operations are carried out in a single station, the overall design mostly depends on the component sizes. An outline of all these forming dies is given in Chapter 2. However, certain specific examples are worked out in this chapter, particularly of bending, drawing and combination dies.

5.2   CLASSIFICATION OF BENDING AND OTHER FORMING DIES

Following are the dies dealt with in this chapter:

5.2.1 Bending Dies

In this die, the strip is bent into a definite required shape along a straight edge. This feature of straight edge differentiates the bending dies from the forming dies where the forming is done along a curved edge. The process of bending is done by carrying out the deformations in the plastic zone of the stress–strain relationship of the material under consideration. As the deformation during the plastic zone has to be carried out subsequent to the elastic deformation, the material is deformed 2–5% more than the required angular displacement, so as to compensate for any spring back due to the release of elastic strain. There are other methods like the ironing of the strip to compensate for these phenomena.

There are basically three types of bending dies:
- 'U' or Channel bends
- Wiping bends
- 'V' bends

Examples are worked out for the Channel and Wiping bends in the following pages. Figure 5.1 shows a component which is selected for carrying out channel bending operation. Figures 5.2, 5.3 and 5.4 illustrate the calculations, sectional front view of the bending die and the top view, respectively. Similarly, Fig. 5.5 illustrates a component requiring wiping operation. Figure 5.6 illustrates the calculations followed by the design of a wiping die in Figs. 5.7 and 5.8.

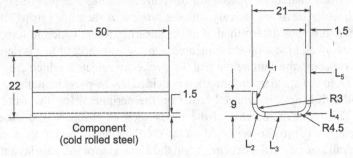

Fig. 5.1 Channel or 'U' Bending of a Component

Length of Bent Part

$$\text{Total Length} = L_1 + L_2 + L_3 + L_4 + L_5$$
$$= (9 - 4.5) + \frac{(2\pi \times 3.75)}{4} + (21 - 4.5 - 4.5)$$
$$+ \frac{(2\pi \times 3.75)}{4} + (22.5 - 4.5)$$
$$= 45.78 \text{ mm}$$

Size of the Blank = 45.78 × 50

Size of the Die = 100 × 100 (Leaving Allowance to Blank Size)

Size of the Punch

Length of the Punch = 21 − (1.1 ×*t*) × 2 (Clearances permitted on either side)
$$= 21 - (1.1 \times 1.5) \times 2$$
$$= 17.7 \text{ mm}$$

Breadth of the Punch = 50 mm

Height of the Punch = Height of Draw + Allowances
$$= 42 \text{ mm}$$

Radius of the Punch = 3 mm

Size of the Die

Radius of the Die = 9 mm

Die Size = 21 × 50

Bending Force for Channel Bending

$$F = \frac{0.67LST^2}{W}$$

L = Length of Bent Part = 45.78 mm

S = Ultimate Tensile Strength (in tonnes/mm^2)

(for Cold Drawn Steel)

= 0.059 tonnes/mm^2

T = Thickness of Strip = 1.5 mm

W = Width of Die = 21 mm

F = 0.67 × 45.78 × 0.059 × 1.5^2

Fig. 5.2 Calculations for Channel Bending of a Component

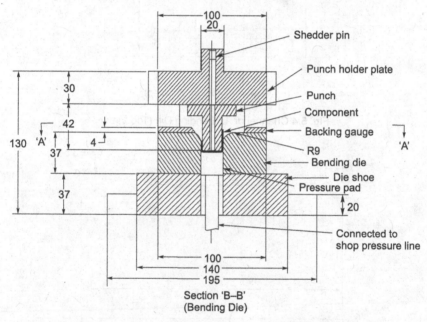

Section 'B–B'
(Bending Die)

Fig. 5.3 Channel or 'U' Bending Die (Cross-Sectional Front View)

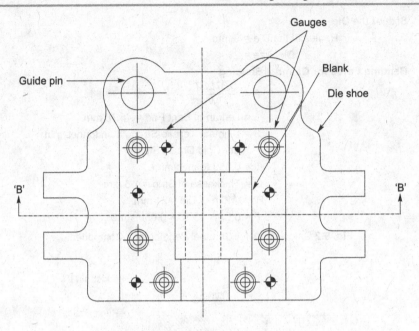

View 'A–A' (before bending)

Fig. 5.4 Channel or 'U' Bending Die (Top View)

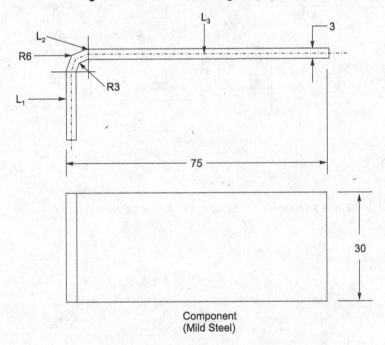

Component
(Mild Steel)

Fig. 5.5 Wiping Bending of a Component

Calculations

$$L = L_1 + L_2 + L_3 \; \Omega \; 95.1 \text{ mm}$$

W = Width of Bend

= Punch Radius + Clearances Die Radius

4(Radius of Component) + 1.07

× Thickness of Blank

+ Die Radius (Inside Radius of Component)

= 12 + (1.07 × 3) + 3

= 12 + 3.21 + 3

= 18.21 mm

Force of Bending F (in tonnes) = $\dfrac{0.33Lt^2 s}{W}$

where

L = Length of Bent Part (in mm)

t = Thickness of Blank (in mm)

S = Ultimate Strength (in tonnes/mm^2)

W = Width of Bend

$F = 0.33 \times 95.1 \times 3^2 \times 0.035 \; 18.21$

= 0.54 tonnes

Fig. 5.6 Calculations for Wiping Bending of a Component

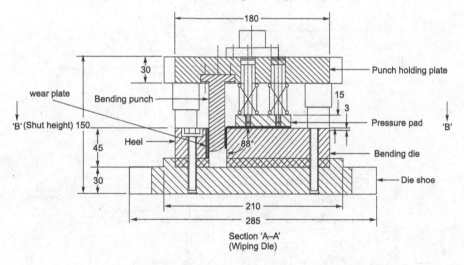

Section 'A–A'
(Wiping Die)

Fig. 5.7 Wiping Bending Die (Cross-Sectional Front View)

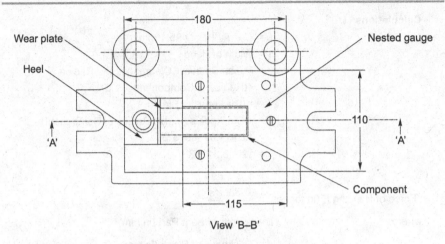

Fig. 5.8 Wiping Bending Die (Top View)

An example is also shown for 'V' bend in Fig. 5.9.

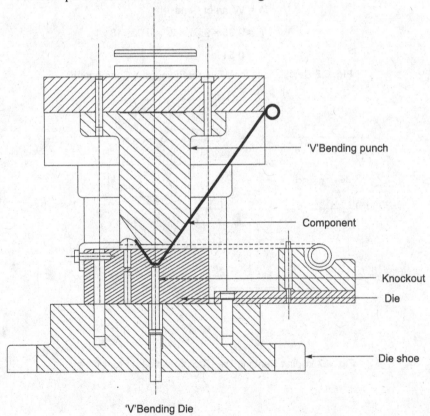

Fig. 5.9 'V' Bending Die (Cross-Sectional Front View)

5.2.2 Drawing Dies

In this die, the blank is drawn in the form of a cup by the displacement of either a punch or a die, depending on the selection and design. Barring the material which is in contact with the bottom of the punch, there is no sizeable stress distribution in this portion; rest of the material of the blank takes the cylindrical shape from its original concentric circular shapes. Force required to draw the material should be optimum and should neither be chosen higher nor lower than that is required, as a higher force may tear the bottom portion of the cup from the cylindrical portion and a lower force may create wrinkles on the surface. By descending order of the magnitude, forces due to blanking are the highest, followed by the drawing forces, and finally the bending forces. The higher drawing forces are due to the fact that the plastic deformations are more severe here than in bending.

There is no stretching involved in normal drawing process, except in certain occasions where it is required, as in the case of extrusion process. Following are the various types of drawing:

(i) *Simple drawing of circular cup without a flange in a single draw:* A typical example is worked out in Fig. 5.10 (component with slanting side), and also in Figs. 5.11, 5.12 and 5.13. The example shown is for a tapered cup.

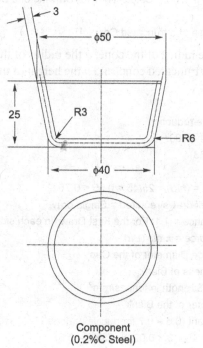

Component
(0.2%C Steel)

Fig. 5.10 Component for Drawing

Calculations

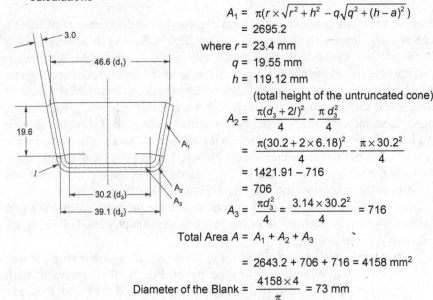

$$A_1 = \pi(r \times \sqrt{r^2 + h^2} - q\sqrt{q^2 + (h-a)^2})$$
$$= 2695.2$$
where $r = 23.4$ mm
$$q = 19.55 \text{ mm}$$
$$h = 119.12 \text{ mm}$$
(total height of the untruncated cone)
$$A_2 = \frac{\pi(d_3 + 2l)^2}{4} - \frac{\pi d_3^2}{4}$$
$$= \frac{\pi(30.2 + 2 \times 6.18)^2}{4} - \frac{\pi \times 30.2^2}{4}$$
$$= 1421.91 - 716$$
$$= 706$$
$$A_3 = \frac{\pi d_3^2}{4} = \frac{3.14 \times 30.2^2}{4} = 716$$

Total Area $A = A_1 + A_2 + A_3$

$$= 2643.2 + 706 + 716 = 4158 \text{ mm}^2$$

Diameter of the Blank $= \dfrac{4158 \times 4}{\pi} = 73$ mm

Fig. 5.11(a) Calculations for Drawing of a Cup

Note: The surface area of a truncated Cone: $\Pi(r\sqrt{r^2 + h^2}) - q\sqrt{q^2 + (h-a)^2}$

where r is the base radius of the cone, q the radius of the truncated cone at the top, a the height of the truncated cone, and h the height of the normal cone without truncation.

Calculations

(a) Percentage reduction
$$= (1 - d/D) \times 100$$
$$= (1 - 45/73)$$
$$= 37.5$$

(b) Draw ratio = $(h/d) = 25/45 = 0.55 < 0.75$

(c) Numbers of draws required = Single draw

(d) Die Clearance = $1.1t$ For the First Draw (in each side) = 3.3 mm

(e) Drawing force = π dts ($D/d - C$)
   d = Average Diameter of the Cup
   r = Thickness of Blank
   s = Yield Strength in tonnes/mm^2
   D = Diameter of the Blank
   C = Constant (0.6 – 0.7 for ductile material)
   $= 3.14 \times 45 \times 3 \times 0.0208 (73/45 - 0.6)$

(f) Radius of Punch and Die = $4 \times t = 12$ mm each.

Fig. 5.11(b) Calculations for Drawing of a Cup (contd.)

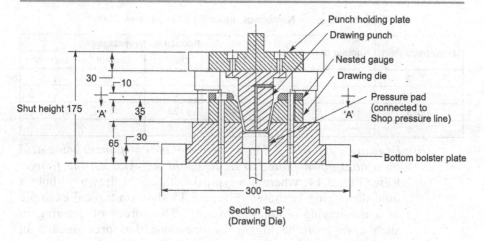

Fig. 5.12 Drawing of a Cup in a Drawing Die (Sectional Front View)

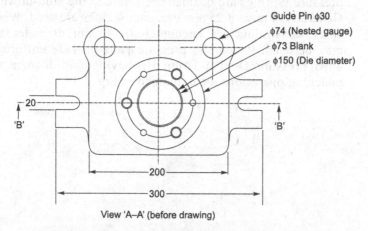

Fig. 5.13 Drawing of a cup in a Drawing Die (Top View)

(*ii*) *Drawing of cup in a number of stages due to large height-to-shell diameter ratio*: Refer Table 5.1 for possible number of draws for a given Draw Ratio. Number of draws can also be decided by the percentage reduction arrived from the formula 100 $(1 - d/D)$. If the said percentage exceeds 50%, then the second draw is considered. However, for practical considerations, even if the percentage exceeds 40%, the second draw is recommended. Drawing a cup in an inverted punch with die carries out the reciprocating motion, the punch being stationary. This type of drawing dies are deployed in deep drawing of components requiring high level of concentricity and precision. These are also used in reverse redrawing. In Fig. 5.25, pressure pads are deployed at the die end, so that when the die descends, the same has to overcome the drawing force as well as the pressure developed by the pad.

Table 5.1 Possible Number of Draws for a Given Draw Ratio

Draw Ratio (*h/d*)	Number of Reductions	Reduction Percentage (%)			
		Draw I	Draw II	Draw III	Draw IV
≤ 0.75	1	40			
0.75–1.5	2	40	25		
1.5–3	3	40	25	15	
3–4.5	4	40	26	15	10

(*iii*) *Drawing a cup with a flange portion of the component being held by a hold-down plate, so as to avoid wrinkles on the flange*: Refer Fig. 5.14, wherein a typical component drawn without a hold-down ring is shown. Figure 5.15 shows a typical example of a die having a hold-down ring. The forces of drawing in such cases will be higher by one-third the force needed in normal drawing, as the punch has to overcome the hold-down pressure. Springs are decided according to the hold-down force. Generally, 4 nos. of 25 mm diameter springs are used. Whenever space constraints are encountered, 20 mm diameter springs may be used. Pneumatic pressure pads provide uniform hold-down pressure. Double-action dies having hold-down as well as punching operations are explained in Fig. 5.16.

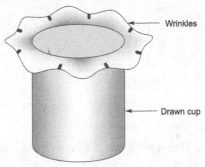

Fig. 5.14 Component with Flange Drawn without a Hold-Down Ring

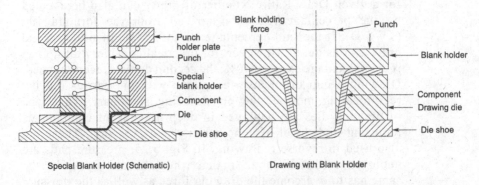

Special Blank Holder (Schematic)   Drawing with Blank Holder

Fig. 5.15 Drawing with a Blank Holder-Specific Example

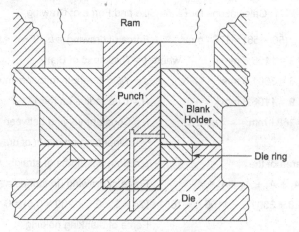

A typical double-action cylindrical draw die

Fig. 5.16 Double-action Draw Die with a Blank Holder

(*iv*) *Drawing a rectangular or square cup*: Here, the area calculation of a blank is different from the usual calculation of blank diameter in drawing of a circular cup. A typical example is shown in Fig. 5.17 and 5.18. Basically, the component is divided into a number of segments and the individual areas are calculated; the total area will be the sum of the individual areas. Certain percentage of allowance has to be provided for in square cups to allow for wrinkles formed at the corners and also for the final trimming of the edges.

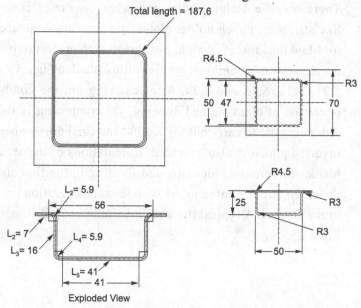

Exploded View

Fig. 5.17. Square Component Drawn with Flanges all round

Calculations for Blank Size and Force of Drawing

$A_1 = (70 \times 70) - (56 \times 56) = 1764$ mm^2 Force of drawing $= ts\ (2 \times \pi \times R \times k_a + L \times k_b)$

$A_2 = 187.6 \times 5.9 = 1106.8$ mm^2 where t = Thickness of Blank

$A_3 = 16 \times 187.6 = 3001.6$ mm^2 s = Yield Strength in tonnes/mm^2

$A_4 = 187.6 \times 5.9 = 1106.8$ mm^2 R = Cup radius (in mm)

$A_5 = 41 \times 41 = 1681$ mm^2 k = Constant varying between 0.5 and 2.0

                                            depending on depth of draw

Total surface area of the L = Sum of lengths of straight sides

component $= A_1 + A_2 + A_3 + A_4 + A_5$ Force of drawing (in tonnes $= 1.5 \times 0.037$

$= 1764 + 1106.8 + 3001.6 + 1106.8 + 1681$ $(2\pi \times 3 \times 0.6 + 200 \times 0.5) = 6.1$ tonnes

$= 8660.2$ mm^2 Force of blanking holding

Size of the blank = 93.06 mm^2 $= 1/3 \times$ drawing force = 2 tonnes

Providing allowances for trimming, etc.,   Total force = 8.1 tonnes.

Size of square blank = 94 mm

Fig. 5.18 Calculations for Square Component Drawn with Flanges all round

(v) *Drawing a cup in a combination die*: Combination die is one in which blanking and drawing take place in a single station. The two operations take place with fraction-of-a-second time-lag with blanking operation preceding the forming operation. Therefore, the design of dies and other associated components like strippers, punch holder plates, die shoe are done according to blanking forces which are greater than forming forces. An example of a combination die is illustrated in Figs. 5.19, 5.20, 5.21, 5.22, 5.23 and 5.24. After carrying out the combination operation of blanking and drawing, the component is shifted to redrawing die to carry out the second and final draw, wherein the inverted punch is also provided. Calculations of the area of the blank, the forces of blanking and forming in the first stage and the calculations related to the second-stage operation are shown in the figures. A typical die with inverted punch is illustrated in Fig 5.25.

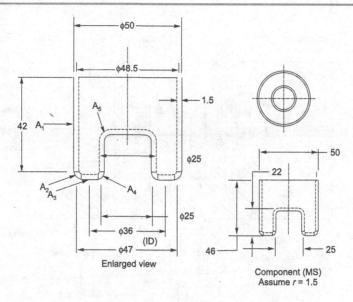

Fig. 5.19 Component for Combination Die

Calculations

A_1 = 3.14 2 × 48.5 × 42 = 6400.25

A_2 = 3.142 × 47 × 3.5 = 516.85

A_3 = (3.142 × 36 × 36)/4 = 1018

A_4 = 3.142 × 25 × 3.5 = 275

For calculation of area A_5 determine the blank dia. of inner cup by using the formula:

Force of blanking = $\pi Dt \times f_s$

D = Diameter of the blank

t = Thickness of the blank

f_s = Shear strength of the material

= π114 × 1.5 × 0.036

= 19.34 tonnes

Force of drawing = $\pi dt\, s(\text{Did} - 0.6)$

$$D = \sqrt{(d-2r)^2 + 4d(h-r) + 2\pi r(d - 0.7r)} = 48.4$$

A_5 = (3.142 × D × D)/4 = 1840

Total Area = $A_1 + A_2 + A_3 + A_4 + A_5$ = 10050.1

Calculated diameter of the blank

= 113.11 (say 114 mm)

Percentage reduction

= 100(1 − (*ID* of drawn shell/OD

of blank) = 100(1 − 47/114) = 58.8%

Since the value is above 40% two drawn are considered. Let the diameter of the cup after the first draw is 75 mm. Then the height of the cup drawn will be

S = Yield Stress (in tonnes/mm²)

d = Diameter of the shell (in mm)

t = Thickness of the blank

0.6 = Constant due to friction

for ductile material

= π × 75 × 1.5 × 0.02 (114/75 − 0.6)

= 17.25 tonnes

Since the blanking operation precedes drawing, the die is designed for a blanking force of 19.34 tonnes.

$$D = \sqrt{d_1^2 + 4d_1 h_1}$$

h_1 = 24.6 mm

Where h_1 and d_1 are the height and diameter of the cup after the First Draw

Calculation for determination of second-stage drawing Initial drawing force

= $\pi d_1 ts(D_1 /d_1 - 0.6)$

= 3.14 × 50 × 1.5 × 0.02 (75/50 − 0.6)

= 4.24 tonnes.

Fig. 5.20 Component for Combination Die–Calculation of Forces

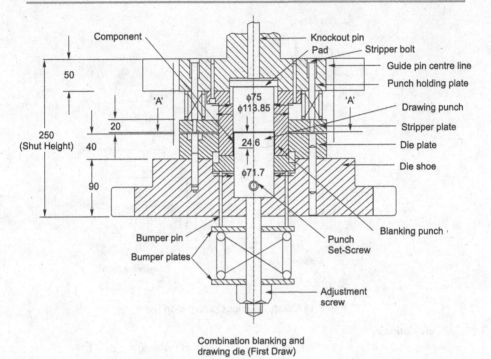

Combination blanking and
drawing die (First Draw)

Fig. 5.21 Combination (Front Sectional View)

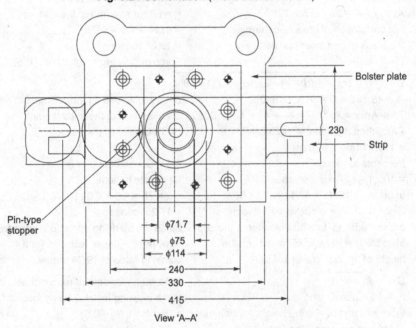

View 'A–A'

Fig. 5.22 Combination Die (Top View)

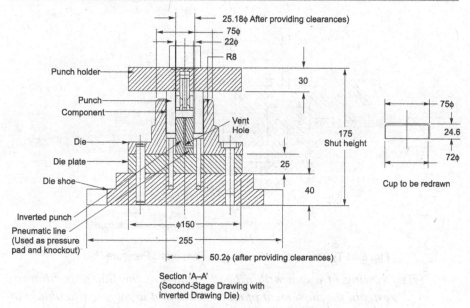

Fig. 5.23 Drawing Die with Inverted Punch (Sectional Front View)

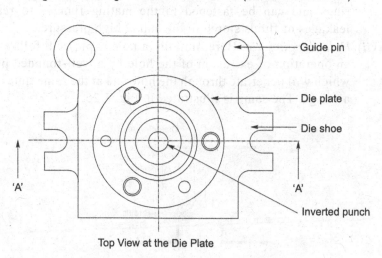

Top View at the Die Plate

Fig. 5.24 Drawing Die with Inverted Punch (Top View)

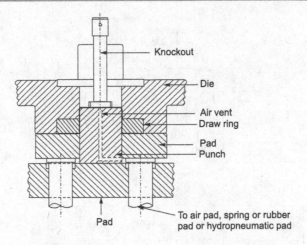

Fig. 5.25 Typical Inverted Punch with Pneumatic Pressure Pad

(*vi*) *Drawing of a cup with a bead located at the flange portion to restrain the movement of material in that area, so as to obtain a wrinkle-free flange*: Beads may be used for providing elastomer rings and can be fastened to the mating flanges to restrain leakages of fluids stored in the cup at high pressure.

(*vii*) *Hole flanging die*: Here, initially a hole is pierced, followed by an operation of extrusion of the hole by a well-rounded punch, which will penetrate through the hole and at the same time create a flange. The same is illustrated in Fig. 5.26.

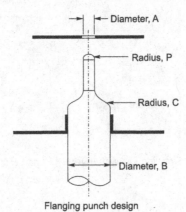

Flanging punch design

Fig. 5.26 Flanging Process through Extrusion of a Pilot Hole

Variables affecting drawing operation

Following are the variables which affect the drawing operation:

(*i*) *Material softness or hardness*: Ductile material, which has a low yield point such as mild steel, has good draw-ability characteristics than a high-carbon or a stainless steel material.

(*ii*) *Material thickness*: Lower material thickness enables easy drawing ability. Material up to 1.5 mm thickness are generally used for deep drawing purposes, rather than materials of 3 mm thickness which are used in shallow cups, sinks, etc.

(*iii*) *Draw clearances between the die and the punch*: Table 5.2 shows the various clearances allowed in drawing operations. These are the optimum clearances given. An increase in clearance will cause wrinkles and surface unevenness. Too close clearance will result in seizure due to excessive heat produced in drawing.

Table 5.2 Clearances Generally Provided in Drawing
Operations between Die and Punch

Blank thickness (mm)	First Draw	Redraws	Sizing Draw (used for fine blanking)
≤ 0.375	$1.07 - 1.09t$	$1.08 - 1.1t$	$1.04-1.05t$
0.4–1.25	$1.08 - 1.1t$	$1.09 - 1.12t$	$1.05-1.06t$
1.25–3	$1.1 - 1.12t$	$1.12 - 1.14t$	$1.07-1.09t$
≥ 3.5	$1.12 - 1.14t$	$1.15 - 1.2t$	$1.08-1.1t$

(*iv*) *Lubricants used*: Chlorinated oils, which include mineral oils that can be formed as emulsions, have the capability to enhance the die life. Other types of oils such as sulphurised oils (mineral oils containing) sulphur are used whenever vapour degreasing is the main requirement. Fatty oils, soap–fat compounds, dry film soaps, mineral oils are also used whenever properties such as vapour degreasing, prolonging die life, rust prevention, etc. are required.

(*v*) *Punch and die radius*: These variables have significant effects on the drawing process. If the draw radius of a die is too large, the metal will flow easily causing wrinkles, and if it is sharp, then excessive thinning may result. Thus, it has been a general rule that the draw radius should be 4 times the stock thickness. Larger radius such as 6 – 8 times can be decided if the stock thickness is considerable.

In the case of punch radius, the same is decided as per the component drawings and the radius required in the specification of the product. However, this rule is applicable for the first draw only, and for the components requiring more than one draw, a radius of 4 times the stock thickness can be selected for the first draw, and for subsequent draws, radius is progressively reduced. A rough formula for deciding the die radius would be Radius of the die = $0.8\,(D - d)T$; where D is the blank diameter, d the shell diameter and T the thickness of the stock.

5.2.3 Other Forming Dies

(a) **Curling dies:** Curled strips are commonly seen in hinges and at the edges of drawn cups. This is basically a forming process, wherein the strip is curled in a circular fashion for a specific purpose. In some cases, the curled strip may have a wire inside as a reinforcement material, to provide strength to the formed portion of the strip. Figure 5.27 shows how progressively curling is done. The diameter of the curl should be more than twice the thickness of the stock. The process of curling in a die is illustrated in Fig. 5.28. The bottom die is supported by a pressure pad/ knockout rod and the drawn cup is placed on it. The upper moving punch has the profile of the curl to be achieved. The upper punch moves down the die bottoms down, displacing the knockout rod down. Subsequently, the curling is effected. After the curling operation is complete, the pressure pad moves up the die and the knockout rod pushes out the component, so as to facilitate its removal. The process does not need as much force as in the case of bending.

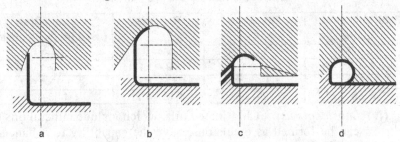

Fig. 5.27: Curling Operation—Progressive Method

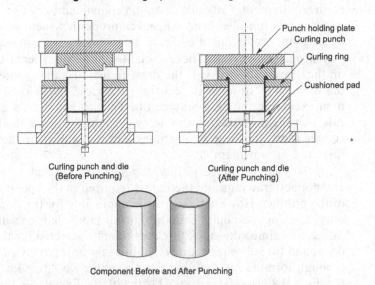

Curling punch and die
(Before Punching)

Curling punch and die
(After Punching)

Punch holding plate
Curling punch
Curling ring
Cushioned pad

Component Before and After Punching

Fig. 5.28: Typical Example of a Curling Die

(*b*) **Swaging dies:** In this operation, one end of the component, say a cylindrical cup, is reduced in cross-sectional size, whereby the length at that portion is increased. Figure 5.29 shows a component as well as a swaging die in operation. Here, the die moves up and down and the internal profile of the die is similar to the component being swaged. Knockout rod is provided at the bottom for displacing the finished component.

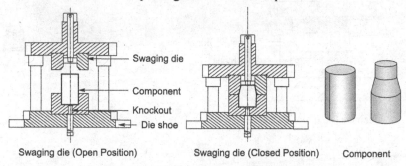

Fig. 5.29: Typical Example of a Swaging Die

(*c*) **Bulging dies:** This is just a reverse of swaging. Using rubber foam or fluids such as oil or water, which can take any shape, the bulging operation is performed. Figure 5.30 illustrates the principle of operation of a bulging die. The descending die enables the top portion of the component to bulge out according to the internal profiles of the die and the punch. The rubber being easily formable, takes the shape of the bulged component. Without this intermediate medium, *viz.* rubber, the component may collapse inside. Overall length of the bulged material reduces in this process.

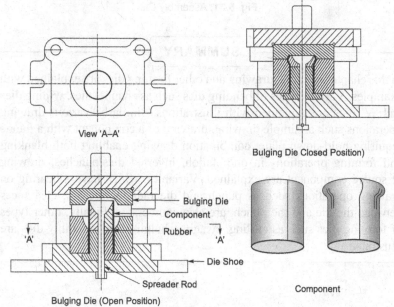

Fig. 5.30: Bulging Die

(*d*) **Assembly dies:** In this die, riveting of components is done, and is applicable in soft materials like aluminium. The operation of riveting a main component with two studs is explained in the Fig. 5.31.

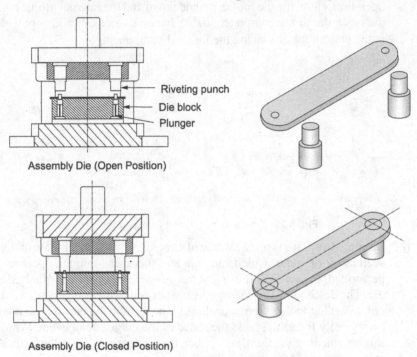

Assembly Die (Open Position)

Riveting punch
Die block
Plunger

Assembly Die (Closed Position)

Fig. 5.31: Assembly Die

SUMMARY

In this chapter bending, drawing and other forming dies are explained with examples. Various types of bending dies such as channel dies, wiping dies and 'V' dies are explained with illustrations. Similarly, various drawing operations such as simple drawing, drawing of a component with a flange requiring hold-down plate, combination drawing enabling both blanking and forming operations in one station, inverted dies/punches, drawing of square component are explained. Variables, which have a bearing on drawing operations such as punch and die radii, lubricants, clearances between the die and the punch, are also enumerated. Finally, other types of forming dies such as curling, swaging, bulging and assembly dies are illustrated.

REVIEW QUESTIONS

1. Distinguish between forming and drawing with reference to area of the blank and the force required for drawing.
2. What are the steps needed to manufacture a stainless steel spoon?
3. Draw out a procedure and show by sketches the design of blanking and drawing dies for the following items:
   - Bath tub
   - Gas stove
   - Stainless steel sink.
4. How to prevent wrinkles on the surface of the flange of a drawn cup?
5. Explain the process of Swaging. Write an expression for the thickness of the swaged portion of a cup in relation to the thickness of the original cup.

ANNEXURE TO BENDING, DRAWING AND FORMING DIES (CHAPTER 5)

COMPUTER MODEL–PROCESS OF MAKING A TIFFIN BOX

A tiffin box is a common consumer item used by millions in the country. Although this item appears to be simple, it is an excellent example for any student of press tools to study how the operations such as blanking, drawing, folding of corners, and bulging of the sides are done to form the body of the container. Subsequently, the lid portion is blanked, drawn, embossed and curled at the edges. Thus, both these components can make a perfect sliding fit.

Figures A5.1 to A5.8 illustrate the computer model process of making the said item. Blanking and drawing can be done in a single combination die. Subsequently, bulging can be carried out as in the case of the bottom half. In the case of the top half, embossing is done simultaneously with drawing. Curling is done at the edges to improve the appearance and to provide certain strength to the container.

Computer softwares such as AutoCad/ProE provide the facility of calculating the overall surface area of the component, which facilitates the calculation of the blank size. In this specific example, while the diameter of the container is 125 mm, the blank size is calculated to be 271 mm, providing for 0.3% allowance in overall diameter. Thickness of the strip is assumed to be 0.75 mm.

Fig. A5.1 Blank Cut from the Strip

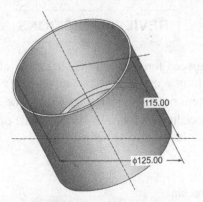

Fig. A5.2 Cup Drawn from the Blank in a Combination Die

Fig. A5.3 Bottom Half of the Container after Performing Operations Like Folding of the
Edges and Bulging of the Cylindrical Sides

Fig. A5.4 Top Half of the Container after Performing Operations Like Embossing of the Top
Surface and Curling of the Edges

Fig. A5.2 Assembled View of the Container (Tiffing Box)

Fig. A5.6 Exploded View of the Container (Tiffin Box)

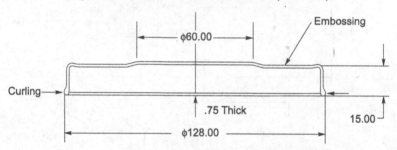

Fig. A5.7 Dimensioned Drawing of the Lid

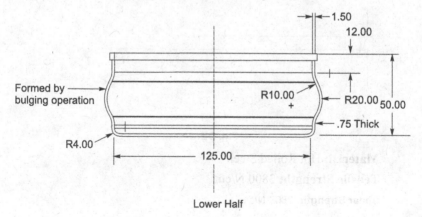

Fig. A5.8 Dimensioned Drawing of the Lower Half

Design Exercises for Press Tools

1. **Design a Combination Piercing and Bending Die for the workpiece shown in Fig. DE. 1.**

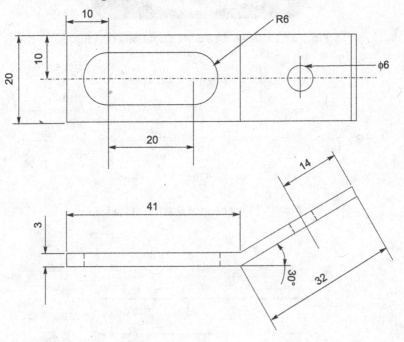

Fig. DE. 1

Material: Hot Rolled Steel

Tensile Strength: 3800 N/cm^2

Shear Strength: 2800 N/cm^2

Design of Jigs, Fixtures and Press Tools, First Edition. K. Venkataraman.
© K. Venkataraman 2015. Published by Athena Academic Ltd and John Wiley & Sons Ltd.

2. **Design a Forming Die from an aluminium strip for the component shown in Fig. DE. 2.**

   Material: Aluminium alloy

   Tensile Strength: 15,000 N/cm^2

   Yield Strength: 10,000 N/cm^2

3. **Sketch a Simple Progressive Die block to blank the workpiece shown in Fig. DE.3, Show:**

   (a) Die block thickness

   (b) Distance between the die opening and the outside edge of the die block

   (c) Location of holes for cap screws

   (d) Location of dowel holes

   (e) Type of tool steel and hardness after heat treatment.

   Material: Cooper Alloy

   Shear Strength: 16,500 N/cm^2

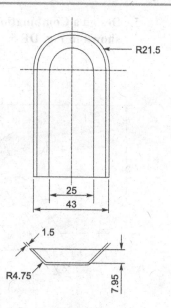

Fig. DE. 2

4. **Design a Combination Piercing and Forming Die for the workpiece shown in Fig. DE.4.**

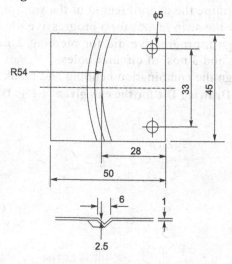

Fig. DE.4

Material: Low-Carbon Steel (annealed)

Tensile Strength: 31,000 N/cm^2

Shear Strength: 31,000 N/cm^2

Yield Strength: 17,250 N/cm^2

5. **Design a Combination Piercing and Bending Die for the component shown in Fig. DE.5.**

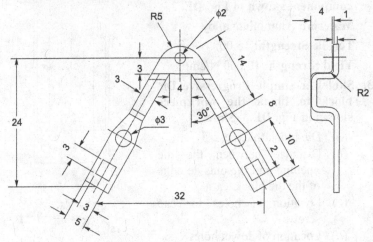

Fig. DE.5

Material: Monel

Shear Strength: 29,580 N/cm^2

Tensile Strength: 48,000 N/cm^2

Hint:

1. Determine the overall length of the workpiece.
2. Draw the strip layout for a progressive die.
3. Design a progressive die for piercing 2 nos. of rectangular holes and 3 nos. of circular holes.
4. Design the combination bending and blanking die.

6. **Design a Drawing Die for the cup given in Fig. DE.6.**

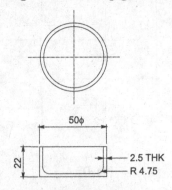

Fig. DE.6.

Material: Chromium Molybdenum Steel

Yield Strength: 41,500 N/cm^2

Appendix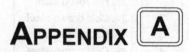

Properties of Materials

Table A.1 Ultimate Shear Strength of Various Materials

Sl. no.	Materials	Shear strength (N/mm^2)
1	Aluminium, cast	92
2	Aluminium, soft sheet	115
3	Aluminium, hard sheet	192
4	Brass, cast	276
5	Brass, soft sheet	230
6	Brass, hard sheet	332
7	Bronze, gunmetal	244
8	Bronze, Phosphor	305
9	Copper, cast	221
10	Copper, rolled	248
11	Cupro Nickel	305
12	Duralumin, treated	267
13	Duralumin, treated and cold rolled	305
14	Steel casting	460
15	Boiler plate	460
16	Angle iron	460
17	Tool steel (drill bit)	620
18	Silicon steel	496
19	Stainless steel	534
20	0.25% C, Mild steel	381
21	1%C, Steel	648
22	Tin-rolled sheet	38
23	Zinc-rolled sheet	138

Design of Jigs, Fixtures and Press Tools, First Edition. K. Venkataraman.
© K. Venkataraman 2015. Published by Athena Academic Ltd and John Wiley & Sons Ltd.

Table A.2 Tensile Properties of some of the Materials

Sl. no.	Materials	Tensile strength (N/mm^2)	Yield strength (N/mm^2)	Elongation in 50 mm specimen (%)	Reduction in area (%)
1	0.2% Hot rolled steel	375	204	25	50
2	0.2% Cold drawn steel	416	347	15	40
3	0.4% Hot rolled steel	518	286	18	40
4	0.4% Cold drawn steel	579	484	12	35
5	0.95% Hot rolled steel	818	450	10	25
6	AISI 301 stainless steel	681	238	50	60
7	Yellow Brass	416	204	62	—
8	Deoxidized Copper	218	68	45	—
9	Magnesium	204	170	4–16	—

Table A.3 Typical Materials Used in Blanking and Piercing Dies

Sl. no.	Material to be blanked	Material of the die punch	
		Upto 100,000 parts	Above 100,000 parts
1.	Aluminium, Copper and Magnesium Alloys	Tool Steel; 0.9%C, 1%Mn, 0.5%Cr, 0.5% Mo Tool Steel: 1.5%C, 12%Cr, 1% Mo	Tool Steel: 1.5%C; 12%Cr, 1% Mo, 1% V; Carbides
2.	Carbon and Alloy Steel upto 0.7%C and Ferritic Stainless Steel	Tool Steel: 0.9%C, 1%Mn, 0.5%Cr, 0.5% Mo Tool Steel: 1%C, 5%Cr, 1%Mo	Tool Steel: 1.5%C, 12%Cr, 1%Mo, 1% V; Carbides
3.	Stainless Steel: austenitic variety	Tool Steel: 0.9%C, 1% Mn, 0.5%Cr, 0.5%Mo Tool Steel: 1%C; 5%Cr; 1%Mo Tool Steel: 1.5%C, 12%Cr, 1%Mo, 1%, V	Tool Steel: 1.5%C, 12%Cr; 1% Mo; 1% V; Carbides
4.	Spring Steel	Tool Steel: 1.5%C, 12%Cr, 1%Mo, 1% V Tool Steel; 1%C, 5%Cr, 1%Mo	–do–
5.	Electrical Steel: Transformer grade	Tool Steel: 1.5%C, 12%Cr, 1%Mo, 1% V Tool Steel: 1%C, 5%Cr, 1%Mo	–do–

Contd...

| 6. | Plastic Sheets: without reinforcements | Tool Steel: 0.9%C, 1%Mn, 0.5%Cr, 0.5%Mo | Tool Steel: 1.5%C, 12%Cr, 1%Mo, 1% V |
| 7. | Plastic Sheets: with reinforcements | Tool Steel: –1%C, 5%Cr, 1%Mo | Tool Steel: 1.5%C, 12%Cr, 1%Mo, 1% V Carbides |

Table A.4 Typical Materials Used in Forming Dies of Mild Severity

| Sl. no. | Materials to be Blanked | Materials of the die or punch | |
		Upto 100,000 Parts	Above 100,000 Parts
1	Aluminium, Copper and Magnesium Alloys	Alloy Cast Iron	Alloy Cast Iron; Tool Steel: %C, 5%Cr, 1%Mo
2	Low-Carbon Steel	Alloy Cast Iron	Alloy Cast Iron; Nitrided Tool Steel: 1.5%C, 12%Cr, 1%Mo, 1%V
3	Stainless Steel: type 300 to ¼ hard	Alloy Cast Iron	Tool Steel: 1.5%C, 12%Cr, 1%Mo,1% V; Nitrided Tool Steel: 1%C, 5%Cr, 1%Mo
4	Heat-resistant Steel	Alloy Cast Iron	Nitrided Tool Steel: 1.5%C, 12%Cr, 1%Mo, 1%V

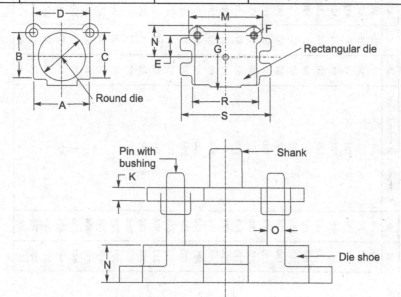

Fig. A.1 Dimensions of a Standard Die Set

Table A.5 Suggested Dimensions of a Standard Die Set (Refer Fig. A.1)

A	B	Diameter of round dies	J	K	C	D	E	F	G	M	N	O	R	S
76	76	76	32	25	86	83	44	25	100	136	71	19	133	171
101	76	82	38	25	87	82	44	30	101	146	76	22	158	203
101	101	101	44	31	113	82	55	30	127	146	87	22	158	196
127	95	101	44	31	109	101	47	31	123	171	81	22	177	222
127	127	127	50	44	138	101	68	31	152	168	101	22	177	222
165	107	127	69	57	120	130	53	34	136	203	90	25	219	285
165	152	165	63	57	165	130	83	34	182	203	117	25	222	285
190	133	146	63	57	146	165	69	34	161	238	109	25	247	311
190	177	190	63	57	190	165	93	34	206	238	130	25	247	311
215	158	177	69	57	171	196	79	34	187	269	115	25	273	336
215	203	215	63	57	215	196	106	34	233	269	142	25	273	336
254	171	–	69	57	185	234	90	36	203	307	128	28	311	377
254	254	254	69	57	269	260	133	41	292	346	176	31	311	377
285	190	203	82	57	206	260	101	41	228	346	144	31	333	409
285	234	254	82	57	250	260	123	41	273	346	166	31	342	409
317	254	273	82	57	273	355	136	50	295	457	190	38	374	438
317	304	317	82	57	323	355	161	50	346	457	215	38	374	438
355	260	285	82	57	285	355	142	50	307	457	196	38	425	495
355	355	355	82	57	374	355	187	50	396	457	241	38	425	495
381	228	–	63	50	247	355	127	50	269	457	180	38	438	504
444	254	–	57	50	273	425	139	50	295	527	193	38	501	565
444	304	–	76	57	323	425	158	50	346	527	212	38	501	565
444	355	381	76	57	374	425	184	50	396	527	238	38	492	565
444	406	431	76	57	425	425	209	50	447	527	263	38	501	565
508	304	–	76	50	323	425	158	50	346	527	212	38	565	631
508	355	–	76	50	374	425	184	50	396	527	238	38	565	631
571	406	–	76	50	425	425	209	50	447	527	263	38	628	695

APPENDIX

Drawing Speeds and Lubricants

Table B.1 Drawing Speeds in Various Materials

Materials	Drawing Speed (m/min)
Aluminium	46 – 53
Brass	53 – 61
Copper	38 – 46
Steel	5.5 – 15.2
Stainless steel	9 – 12
Zinc	38 – 46

Table B.2 Lubricants Commonly Used in Drawing (Water-based Lubricants)

Composition of Lubricant	Properties		
	As Water-based Cleaners	As Degreasers	As Rust-preventing agent
For low severity (less than 10%): water-based emulsion of soluble oil or wax	Very Good	Good	Fair
For medium severity (between 11–20%): water-based emulsion of soap, water-based emulsion of heavy duty soluble oil (contains additives like sulphur or chlorine)	Very Good Very Good	Very Poor Good	Fair Fair
For high severity (21–40%): water-based emulsion of soluble oil with high content of sulphurised or chlorinated additives; or soap–fat paste	Very Good Fair	Good Poor	Fair to Poor Fair
For maximum severity (above 40%): pigmented soap–fat paste diluted with water; or dry soap or wax diluted in water with fillers as borax	Poor Good	Very Poor Very Poor	Good Good

Table B.3 Lubricants Commonly Used in Drawing (Oil-based Lubricants)

Composition of Lubricants	Properties		
	As water-base cleaners	As Degreasers	As rust preventing agent
For low severity (less than 10%): mineral oil	Good	Very Good	Fair
For medium severity (11–20%): mineral oil plus fatty oils; or mineral oil with sulphurised or chlorinated oil	Good Good to Fair	Very Good Good	Fair Fair to Poor
For high severity (21–40%): mineral oil plus non-emulsifiable chlorinated oil; or mineral oil with emulsifiable chlorinated oil; or fatty oils	Poor Good Fair	Good Good Fair	Very Poor Very Poor Fair
For maximum severity (above 40%): concentrated chlorinated oil (non-emulsifiable); concentrated chlorinated oil (emulsifiable); concentrated sulfo-chlorinated oil (non-emulsifiable); concentrated sulfo-chlorinated oil (emulsifiable)	Very Poor Good Very Poor Good	Fair Fair Fair Fair	Very Poor Very Poor Poor Poor

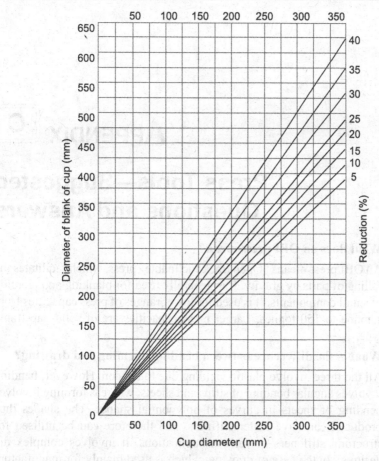

Fig. B.1 Plot Showing the Relationship Between Cup Diameter, Blank Diameter and Percentage Reduction

APPENDIX C

Press Tools—Suggested Questions and Answers

Q. 1. **What does an OBI press mean?**

Ans. An OBI press means "Open Back Inclinable" press, which facilitates the falling of parts by gravity. It is generally used for blanking and piercing of small components. It is used for lower ranges of press capacities from 1 tonne to 150 tonnes. Larger press capacities are of solid gap-frame type.

Q. 2. **What is the difference between bending, forming and drawing?**

Ans. All the three involve plastic yielding due to tension. However, bending involves angular bending of strips and sheets; whereas forming involves bending of sheets in curves of polynomial shapes. The shapes thus produced can have better stiffness, and therefore can be utilised for structural stiffeners in various applications. It involves complex die designs. In the case of drawing, which is used mainly for manufacture of tubes, cups, etc., tensile yielding is effected only in the drawn tube or cylindrical portion. The bottom portion is not strained. However, the top flange is subjected to compressive stresses and hoop strains.

Q. 3. **What is the function of draw die ring?**

Ans. The function of draw die ring is to hold the flange portion, so as to effect the drawing of the cup or tube. This eliminates the formation of wrinkles at the flanges, as it forms a grip while drawing.

Q. 4. **What is a disappearing pin? Where is it used?**

Ans. A disappearing pin is used to act as a stopper in blanking jobs. The stopper works to stop the strip movement while the strip moves from one station to another, as in the case of progressive dies. The stopper projects at the recessed portion of the strip and dips itself while the strip moves.

Q. 5. **What are the methods of reducing spring back?**

Ans. 'Spring back' can be reduced by ironing, by corner setting and by having negative camber at the bottom of the die.

Design of Jigs, Fixtures and Press Tools, First Edition. K. Venkataraman.
© K. Venkataraman 2015. Published by Athena Academic Ltd and John Wiley & Sons Ltd.

Q. 6.   What is meant by air-bending?

Ans. Air-bending means bending the strip by a punch using the span of the die block, without the strip actually making contact with the die surface.

Q. 7.   What is the purpose of primary stops?

Ans. The purpose of primary stops is to stop the strip in the beginning of the operation in a progressive die.

Q. 8.   What is meant by curling?

Ans. Curling is the process of forming the strip end in a circular fashion, so that it forms a complete loop. It is used in hinges and in certain utensils requiring smooth curly edges.

Q. 9.   What is meant by a wiping die?

Ans. It is the bending die wherein the bent portion is perpendicular to the original blank. In other words, the bent segment is vertical and is formed due to vertical movement of wiping die.

Q. 10 In sheet metal piercing and blanking, what is meant by penetration?

Ans. It is the distance travelled by the blanking or piercing die to complete the shearing action on the component, so as to enable the fracture to follow.

Q. 11 What is a trigger stop?

Ans. A trigger stop is a stop which is actuated by spring force, which in turn can be initiated by the operator.

Q. 12. What is an air-cushion? What is the advantage of air-cushion?

Ans. Air-cushion is one in which the pressure pads are provided with air-pressure at their bottom to carry out operations in a compound die.

Q. 13. What is the force required for channel bending?

Ans. Force required for channel bending = $(0.67\ Lt^2\ S)/W$, where L is the length of the bent strip, t, the thickness of the stock, S the ultimate strength of the material used, and W the width of the channel.

Q. 14. How is the press tonnage requirement for a drawing operation determined?

Ans. The force for drawing operation is given by the formula: $\pi t s\ (D/d - C)$, where d is the diameter of the cup (Component), D the blank diameter, S the yield strength of the material and C is a constant varying between 0.6 and 0.7 (depending on the ductility of the material). If draw rings are used an additional 33% should be added to the drawing force calculated earlier. If the drawing is done in a combination die having blanking and subsequently drawing, blanking force calculations will take precedence to drawing forces in deciding the press tonnage, as the drawing operation will be carried out subsequent to the blanking operation due to marginal time-lag between the two.

Q. 15. **What type of press is used for air-bending?**

Ans. Press brakes are used for air-bending. Single-action straight-side eccentric shaft mechanical presses can be employed.

Q. 16. **What is meant by lancing?**

Ans. Lancing is the process of shearing of strip in three sides of a rectangle and bending the strip about one of the sides not sheared. It is used for specific applications.

Q. 17. **What is a combination die?**

Ans. A combination die is one which performs blanking and bending in one station in a single press.

Q. 18. **What is meant by 'shut height' of a press?**

Ans. 'Shut height' is the distance between the top of the top bolster to the bottom of the die shoe when the press tool is in shut position or in closed position.

Q. 19. **What is meant by angular clearance?**

Ans. The cylindrical hole at the blanking and piercing die is not designed to be exactly true. In other words, they are ground conical. The angle subtended by the vertical face of the cylinder varies from 0.5° to 1.5° so that the blank punched is ejected out with ease and the angle enables the adequate clearance for the required eviction of the blank, particularly to account for certain spring back of the blanked component.

Q. 20. **What is meant by 'stretch forming'?**

Ans. The process of elastically stretching a strip by gripping on two edges and applying pressure at the centre in a form required, is known as 'stretch forming'.

Q. 21. **What is meant by a double-pass layout? What are its limitations?**

Ans. It is a strip layout process in which the objects which are required to be blanked are laid out in such a way that they form a zigzag pattern along the width of the strip, when the blanking of the strip is carried out in two passes. This enables better utilisation of strip area, although larger strip widths are needed. The limitations are: (*a*) larger width of strips are required to be slit for such operation to be performed and (*b*) the process requires two passes requiring the same coil to be handled twice.

Q. 22. **What is meant by fool proofing of a die block?**

Ans. Fool proofing of a die block is a unique way of feeding the strip for any press tool operation.

Q. 23. **What is meant by a finger stop?**

Ans. Finger stop is the primary stop provided initially at the start of strip processing, particularly in progressive dies. This enables stopping of the strip in the initial operations in the first few stations before the strip reaches the automatic stop at the end.

Q. 24. How is the stock guided in press tools?

Ans. A stock is guided in press tools by a set of guiding pins located at the edges of strip or by providing channel stripper which will have a recess in the form of channel to guide the strip.

Q. 25. What is the difference between direct and indirect knockout?

Ans. Knockout provided at the punch to push out the stock or blank is known as a direct knockout; whereas in the indirect knockout the same is provided at the die, particularly used in bending and drawing jobs.

Q. 26. What is meant by fine blanking?

Ans. Fine blanking means precise blanking operation requiring closer clearances between the punch and the die, so as to produce burr-free surface of the component. Triple action hydraulic presses requiring holding, drawing and ejection actions are performed to produce such components.

Q. 27. What is the advantage of bottoming dies compared to air dies?

Ans. The advantage of bottoming dies compared to air dies is that spring back of the component can be minimized while bending operations are performed in components such as 'V' bends, channel bends, etc.

Q. 28. What are the advantages of compound die compared to progressive die?

Ans. Compound die performs blanking and piercing operations in single station of the die, whereas progressive die performs in successive stations calling for more die area. Secondly, in the compound die, the concentricity errors of pierced hole and blank are minimal as both these operations take place in the same station. The third advantage is the increased productivity of the die for specific applications.

Q. 29. What is meant by reverse redrawing?

Ans. In the reverse redrawing process, the redrawing operation is performed on the component in a direction opposite to that of drawing.

Q. 30. What is the effect of insufficient clearance between the punch and the die in a cutting operation?

Ans. The effect is excessive shearing force causing undue wear and tear of the punch and die.

Q. 31. What is a double-action press?

Ans. A double-action press is one in which holding and subsequent blanking are done in two strokes of a hydraulic press.

Q. 32. How does a stock stop function in a manually-fed press tool operation?

Ans. Pin-type or cylindrical stops are provided at the end of the blanking operation, and they stop the strip movement by being in contact with the previously sheared edge of the strip. If the movement is needed subsequent to the blanking, the strip is manually moved over the stop up to the distance of "lead" or "advance" of the strip and the sheared edge of the strip is once again allowed to butt against the stop.

Q. 33. What is 'French stop'?

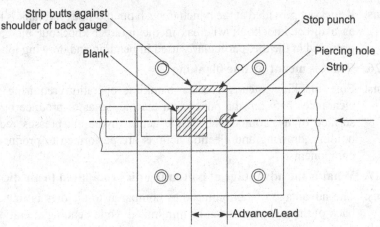

Fig. C.33.1

Ans. French stop is explained in Fig. C.33.1. Basically, the strip having a larger width than that is required is fed to the die. One of the edges butts against the shoulder of the back gauge. At that point of time, the piercing of the hole takes place. Simultaneously, a portion of the strip whose length is equal to the "Advance/ Lead" is also sheared off. This allows for the strip to advance for one pitch, so that the next operation of blanking can take place. The disadvantages are the requirement of larger strip width and the requirement of one additional operation.

Q. 34. What is meant by a compound die?

Ans. Where the piercing and blanking done in single operation and in a single station.

Q. 35. What is the difference between strippers, ejectors and knockouts?

Ans. Strippers are meant to restrict the strip at the die surface itself without being lifted up. Ejectors, otherwise called as pressure pads, are used in drawing and bending jobs, and act as cushion at the die while the punch moves up or down, and enable ejection of the component after the forming operation is completed. Knockouts are specific devices identified and fitted with punches, and perform the operation of pushing out the blank after the shearing operation is over.

Q. 36. What is the difference between coining and embossing?

Ans. Both are same with respect to plastic yielding of the surface of the blank to create desired imprints on the surface. However, in the case of coining, the process is done on both sides with retainer rings provided on the edges to contain the shape. Further, the squeezing needed in coining is enormous (compressive force), so as to create projections on both sides of the coin. In embossing, projections are created in one side only by means of plastic indentation of the punch.

Q. 37. What is the advantage of air-bend dies?

Ans. It is a simple '*V*' bending operation done without the component touching the bottom of the die.

Q. 38. What is the advantage of reversed redrawing?

Ans. The advantage of reversed redrawing is that it ensures circularity of the drawn cup throughout the length and also ensures axial straightness.

Q. 39. What is meant by centre of pressure?

Ans. Whenever unsymmetrical objects are blanked, the resultant force due to press action on the profile of the object need not act at the centroid or the geometric center. Therefore, if the ram of the press is located at the geometric center, there is a likelyhood of buckling loads induced to the press ram. This is not desirable. It is imperative, therefore, to determine the point at which the resultant force acts, which is otherwise called the center of pressure to locate the ram at that point.

Q. 40. What is the difference between direct and indirect piloting?

Ans. Direct piloting is one which is fixed to the punch for locating pierced holes, whereas indirect piloting registers the hole prior to punching.

Q. 41. What causes the burnished surface (often referred to as cut hand) that appears on the walls of the pierced hole and slug?

Ans. The burnished surface is caused due to very close clearance between the die and the punch (say 1 – 2% of the thickness of the strip/component on either side of the punch).

Q. 42 What is the function of a knockout rod?

Ans. The function of a knockout rod is to facilitate the removal of the blank/component subsequent to the blanking operation.

Q. 43. How do you determine the number of draws required for drawing a cup?

Ans. Determine the percentage reduction by the formula:

- $P = 100 \times (1-d/D)$, where d is the inside diameter of the shell and D the blank diameter. If this exceeds 40%, then more than one draw is required. Alternatively, if the ratio h/d (the height of the drawn cup/ the diameter of the cup) is less than or equal to 0.75 single draw is used. If it exceeds 0.75 but falls less than or equal to 1.5, then two draws are used. If the ratio falls between 1.5 and 2, then three draws are used. If it exceeds 2, but falls less than 3, then four draws are used.

Q. 44. What is the type of force coming on the punch? How are the punches designed?

Ans. The type of forces coming on the punch are mainly compressive forces. Punches are designed based on the profile to be formed. The length of a punch is decided to be the same as that of one of the sides of the blank. In the case of circular punch, the length will be equal to the diameter of the punch.

Q. 45. How do you determine the number of stations required for a progressive die?

Ans. The number of stations required for a progressive die depends on the total number of piercing and blanking operations needed to complete the workpiece.

Q. 46. How will you determine the blank diameter for an axisymmetric drawn cup?

Ans. The blank diameter is given as:

$$D = \sqrt{d^2 + 4dh}, \text{ when } d/r \text{ is 20 or more}$$

$$= \sqrt{\left(d^2 + 4dh\right) - 0.5r}, \text{ when } d/r \text{ is between } 15 - 20$$

$$= \sqrt{\left(d^2 + 4dh\right) - r}, \text{ when } d/r \text{ is between } 10 - 15$$

$$= \sqrt{\left(d - 2r\right)^2 + 4d(h - r) + 2\pi r(d - 0.7r)}, \text{ when } d/r \text{ is below 10}$$

where D is the blank diameter, d the tube OD, h the tube height and r the radius of curvature of the tube.

Q. 47. What is the difference between swaging and bulging?

Ans. Swaging is the process of forming a component at any one of its ends to reduce the cross-sectional area, whereas bulging is done over the entire internal surface of a component either by fluid pressure or by using polyurethane foam.

Q. 48. What is the type of press used for drawing? Why?

Ans. Combination die which does blanking as well as drawing in one operation is used for drawing. This is done to achieve better alignment of the drawn product. Inverted dies are best suited for drawing operations.

Q. 49. What is the difference between parting and cutting?

Ans. Parting means dividing the strip through shearing in the two sides of the punch, whereas cutting is done for trimming the scrap by shearing one side of the strip.

Q. 50. What is the effect of excessive clearance between the punch and die cutting operations?

Ans. The effect is plastically deformed shape of the blank without the edges being straight and burr-free.

Q. 51 What is 'hole-flanging'? Why is it done?

Ans. The process is basically to pierce a hole on a strip and subsequently to extrude into a flanged hole. This is explained through Fig. C. 51.1.

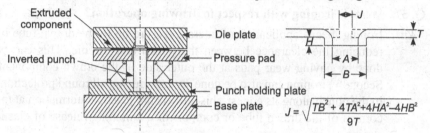

$$J = \sqrt{\frac{TB^2 + 4TA^2 + 4HA^2 - 4HB^2}{9T}}$$

Fig. C. 51.1

Q. 52. What is a triple-action press?

Ans. A triple-action press is one which has the following three features:

    (*a*) Holding or gripping action outer concentric ring

    (*b*) Controlled movement of the punch

    (*c*) Centre of pressure for fine blanking action as well as for ejection of the component.

Q. 53. What are the different methods of attaching pilots to punches?

Ans. The different methods of attaching pilots to punches are:

    (*a*) By directly screwing on to the punch

    (*b*) By having spring cushion (assembled inside the bore made in the punch) on the spindle connecting the pilot and the punch, so as to act as a registering device on to the already pierced hole.

Q. 54. What are the defects in drawn components?

Ans. The defects in drawn components are of the following types:

    (*a*) Spring back

    (*b*) Wrinkles on the flange portion

    (*c*) Excessive thinning of the drawn tube

    (*d*) Burrs on the outer edge of the flange

    (*e*) Cracks formed due to the fact that the formability criteria are exceeded.

Q. 55. What is bend allowance? How is bend allowance calculated?

Ans. Bend allowance is the circumferential distance of the neutral plane of the bent strip after bending is carried out. It is given by

$$B = (A/360) \times 2 \times \pi_i \times 2\ (IR + Kt)$$

where B is the bend allowance, IR the inside radius of the bent strip, K the constant = 0.33 or 0.5 depending on whether IR is less than $2t$ or more than $2t$, and t bending the thickness of the strip.

Q. 56. What is notching?

Ans. To create 'V'-shaped cut profiles at the edges of strips by shearing. An example is keys having serrations either on single side or on both sides.

Q. 57. What is ironing with respect to drawing operation?

Ans. Ironing means application of lateral pressure on to the drawn tube by reducing the clearance between the punch and the die. This can be done by having wear pads at the outer side of the drawn component. Secondly, ironing can also be done by introducing all round projections in the punch along its edges. This will ensure plastic deformation at the corners of the drawn tube or component eliminating release of elastic strain.

Q. 58. What is the difference between a fixed and a floating stripper?

Ans. A fixed stripper is screwed on to the die block, whereas a floating stripper is suspended from a punch holder assembly through a set of spring assemblies, so as to exert optimum pressure on the strip.

Q. 59. What is the function of the 'air-vent' provided on the punch of a drawing die?

Ans. The function of the air-vent is to avoid formation of vacuum between the punch and the blank interface, making it easy for removal.

Q. 60. What is an automatic stop?

Ans. An automatic stop is a spring-actuated stopper located subsequent to the final die station in a progressive die, so as to arrest the movement of the strip when the punching/piercing/blanking operation is to take place. These are generally bought out devices.

Q. 61. What is an inverted die? What is its advantage?

Ans. When the drawing operation is carried out upside down with the punch being at the bottom in an inverted position and the die moves up and down, then it is known as an inverted die. The advantage is that the deep drawn tubes can be produced with much better accuracy and the ejection of the component can be carried out with ease.

Q. 62. How will you determine the centre of pressure for a progressive die?

Ans. The following steps should be carried out to determine the centre of pressure for a progressive die:

(*i*) Initially divide the profile into a number of segments, such as semi-circular curves, straight edges, circular edges, etc.

(*ii*) Assume imaginary $X - Y$ axes adjoining the profile.

(*iii*) Multiply the length of each segment with the co-ordinate distances of the center of gravity of each line segment (length of line segment × co-ordinate distances of the CG of each line segment in x and y direction from the $X - Y$ axes).

(*iv*) Divide the sum arrived as per the procedure explained in step (*iii*) with the sum of the lengths of all the line segments. The net result will be either the distance in *x*-axis of the center of pressure or the distance of the center of pressure in the *y*-axis. The same can be expressed as follows:

$$\overline{X} = (\Sigma x_n l_n) \Sigma l_n$$

$$\overline{Y} = (\Sigma y_n l_n) \Sigma l_n$$

where n = 1, 2, 3, ... and it depends on the number of line segments assumed in the non-symmetric profile.

Here, $\overline{X}$ and $\overline{Y}$ are the co-ordinates of the center of pressure with respect to the assumed *x* and *y* co-ordinates. x^n and y^n are the distances of the mid-point of each line segment along the *x* and *y* axes.

Q. 63. What is the function of draw bead?

Ans. Whenever forming operation is done (as shown in Fig. 62.1), wrinkles are likely to be formed in the component in the area which is not in contact either with the blank holder or with the punch. This is eliminated by providing draw bead of circular shape all round the blank holder. Corresponding recess is provided in the draw ring, so that the sheet is made to pass through the recess while the drawing operation takes place. This results in additional tension being created in the sheet to be drawn in the flange portion eliminating any wrinkles.

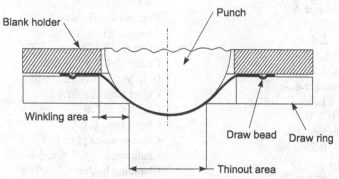

Fig. 62.1 Example of Draw Bead in Drawing Operation

❑❑❑

Index

PART-I

PART-II

REFERENCES

1. American Society of Tool and Manufacturing Engineers, *Fundamentals of Tool Design,* Prentice-Hall of India, New Delhi, 1983.
2. Parsons, S.A.J., *Production Tooling Equipment,* Cleaver–Hume Press Limited, London, UK, 1959.
3. Joshi, P.H., *Jigs and Fixtures*, Tata McGraw-Hill Publishing Company Limited, New Delhi, 2001.
4. Chapman, W.A.J., *Workshop Technology*, Edward Arnold, London, UK, 1975.
5. Donaldson, Cyril, George H. Lecain, and V.C. Goold, *Tool Design,* Tata McGraw-Hill, New Delhi, 1976.
6. ASM, *Metals Hand Book*, 9th International Edition "Forming", Ohio, USA.
7. ASM, *Metals Hand Book*, 9th Edition "Machining", Ohio, USA.
8. Kemster, M.H.A., *An Introduction to Jig and Tool Design*, Hodder & Stongton, UK, 1975.
9. American Society of Tool and Manufacturing Engineers, *Die Design Handbook*, McGraw-Hill Book Company, Inc., USA, 1965.

Design of Jigs, Fixtures and Press Tools, First Edition. K. Venkataraman.
© K. Venkataraman 2015. Published by Athena Academic Ltd and John Wiley & Sons Ltd.

Printed in the United States
By Bookmasters